Christopher Plumb is an independent historian. He has worked as a television and museum consultant and holds a PhD from the University of Manchester.

'From magnificent menageries to an apothecary's pet rodent, Christopher Plumb's book is a fascinating look into the role animals played in eighteenth-century British lives. Full of great primary research into a wealth of interesting records, this is a work to delight the heart of anyone with a love for how the real Georgians lived.'

Lucy Inglis, author of *Georgian London*

'Christopher Plumb's entertaining book fills in the detail of a world only vaguely sensed. It appears that the streets of Georgian London were thronged with exotic animals and Plumb shows that these were more fully a part of the Georgian world than has previously been understood. Exotic animals were commodities to be entertained by and to consume. This book adds not only to our growing understanding of the surprisingly large-scale presence of exotic animals in England since the Renaissance but also to our grasp on the textures of life in the always fascinating streets, inns and drawing rooms of Georgian London.'

John Simons, author of *The Tiger That Swallowed the Boy: Exotic Animals in Victorian England*

THE
GEORGIAN
MENAGERIE

Exotic Animals
～ in ～
Eighteenth-Century
London

CHRISTOPHER PLUMB

Published in 2015 by
I.B.Tauris & Co. Ltd
London • New York
www.ibtauris.com

ISBN: 978 1 78453 084 6
eISBN: 978 0 85773 928 5

A full CIP record for this book is available from the British Library
A full CIP record is available from the Library of Congress

Library of Congress Catalog Card Number: available

Typeset by JCS Publishing Services Ltd, www.jcs-publishing.co.uk
Printed and bound by ScandBook AB, Sweden

Contents

Illustrations

Acknowledgements

I WOULD LIKE TO thank both my editor at I.B.Tauris, Joanna Godfrey, and my agent Kirsty McLachlan for their invaluable advice and assistance; by their hands, an unruly beast of a manuscript was tamed into a book.

The roots of this book lie in a doctoral thesis supervised by Samuel Alberti and the late John Pickstone. As they pruned my academic writing of a surfeit of eighteenth-century animal anecdotes, I was always told with encouragement to 'save it for the book'. In the later stages Helen Rees Leahy and Charlotte Sleigh helped shape the thesis in its final form as doctoral examiners. The doctoral research was supported financially by the Arts and Humanities Research Council UK and the Max-Planck-Gessellschaft, Germany.

This book would have been impossible without access to archival material and primary sources. Mike Rendell very kindly shared the diaries of Richard Hall, his great-great-great-grandfather. Jean Smiter-Brookes, descendant of the bird and animal merchant Joshua Brookes, generously shared the fruits of her research into her family history. I would also like to express sincere gratitude to the staff of the Bodleian Library, the British Library, the British Museum, the London Metropolitan Archives, the National Records Office, the Royal College of Surgeons, London, the National Gallery of Victoria and the Yale Center for British Art.

Some sections of this book have been adapted in part from some of my previously published academic research. Permission to republish has been granted from the following rights holders:

Elsevier for 'The "Electric Stroke" and the "Electric Spark": Anatomists and Eroticism at George Baker's Electric Eel Exhibition in 1776 and 1777', *Endeavour* xxxiv/3 (2010), pp. 87–94.

Maney Publishing for '"In Fact One Cannot See it Without Laughing": The Spectacle of the Kangaroo in London, 1770–1830', *Museum History Journal* iii/1 (2010), pp. 7–32.

Wiley and Sons for '"Strange and Wonderful": Encountering the Elephant in Britain, 1675–1830', *Journal for Eighteenth-Century Studies* xxxiii/4 (2010), pp. 525–43.

University of Virginia Press for 'The Queen's Ass: The Cultural Life of Queen Charlotte's Zebra in Georgian Britain', in Samuel Alberti (ed.), *The Afterlife of Animals* (2011), pp. 17–36.

Prologue

RICHARD HALL (1729–1801) WAS a haberdasher, and had a parrot called Polly. His diaries recorded the times when 'Polly was unwell', when she got better – 'Polly much recovered' – and the time that 'Polly went away'. Hall kept her cage; an inventory in 1794 recorded it as stored in a backroom among clotheshorses and 'sundry boxes'.

Hall was born to a hosier and grew up in an ordinary, even fairly undesirable, part of Georgian London: Red Lion Street, Southwark, a short narrow street lined with inns and taverns. During Richard Hall's childhood, his father's business modestly prospered and so he received some form of formal education more comprehensive than the eighteenth-century parish or dame schools; Richard was literate, had studied French and possessed a few books in Latin.

All this set him in good stead for adult life as one of Georgian London's 'polite and commercial' people. Hall moved from stockings into haberdashery, selling silks, damask, brocades, silver buttons and buckles. Through hard work he built a thriving business and married well. Hall's first wife, Eleanor Seward, came from a small land-owning family; her parents owned a rather run-down early sixteenth-century 'mansion house' in Bengeworth, Worcestershire. Once inherited, this wealth gave the newly married couple a push into the very upper levels of London's middling sort. The children attended boarding schools and the family took up a fine house in Surrey as well as their shop premises at 1 Lower Thames Street, London Bridge. The Halls rented out several farms to tenants so

Richard Hall, although a haberdasher by trade, lived the life of a City gentleman in the 1760s and 1770s.

His family were not, admittedly, ordinary Londoners; there were far greater numbers of Londoners with much less than the Halls. Yet they are quite representative of Georgian London's prosperous mercantile classes: their home was well-appointed and fitted-up genteelly. Hall's clothing purchases reveal a desire to dress to impress his clients – part and parcel of being a City haberdasher. Other Hall family possessions demonstrate a keen awareness of prevailing tastes; they kept goldfish in a 'glass globe', or bowl, that Hall had paid six shillings for. In the 1760s, the period when the Halls had them, goldfish were foreign rarities kept in porcelain or glass bowls; by the end of the eighteenth century goldfish could be found swimming in London's ditches. Alongside the usual trappings of bourgeois comfort – porcelain, silver, mahogany furniture, damask and silk upholstered chairs, Wilton carpets and quality linen – the Halls had possessions such as a walnut-cased harpsichord, preserved moths in glazed cases, as well as a collection of fossils and shells. Richard Hall's habit of scribbling notes down as well as cutting and pasting handbills, newspapers and other clippings into notebooks, in addition to keeping diaries and maintaining his accounts really was a fortuitous one for us. Through this, Hall recorded that he had seen much of London's Georgian menagerie.

Before he married, Hall had already begun his collection of ephemera, including a handbill for a wild animal exhibition at the Talbot Inn in 1752. There he had seen a rhinoceros and a crocodile, and noted so in his diary. Later, Hall and his family attended the Chapel Royal at St James's Palace on 1 April 1768. Here they saw King George III and Queen Charlotte, after which they visited the Queen's zebra, paying two shillings for the pleasure – although it was supposed to be shown to the public as a gracious gratuity. The Halls paid a further three shillings and saw the Queen's elephants. For the Hall family, then, exotic animals were both a familiar part of their domestic everyday life as well as a noteworthy spectacle.

In 1780, after the death of his first wife, the 51-year-old Hall moved to the Cotswolds and made a new life as a gentleman yeoman farmer brewing apple cider, making blackcurrant wine and tending to a garden. His land holdings provided an income for his second wife, Betty Snooke, and their family. Richard's interest in haberdashery waned in middle-age and, although the leasehold on the London Bridge property was kept, his son now dealt with business affairs in his stead. Even away from London, in rural Gloucestershire, foreign birds were part of domestic life; Hall's sister-in-law kept a parrot, and Hall was given a canary as a present by a Mrs Pratt. Hall ordered canary seed especially from London to feed his bird.

In his later years Hall's interest in birds seems to have grown, as he paid for a carpenter to build an aviary in his garden. Hall's sister-in-law owned a parrot too, called 'Poll'; Polly and Poll were not, admittedly, particularly imaginative choices but they did at least follow the prevailing trend in Georgian parrot name-giving. The death of her sister's parrot inspired Betty to pen a droll little elegy which Richard noted in his diary; no doubt it was inspired by the dozens of printed parrot elegies and eulogies printed in fashionable magazines and poetry collections. His diary noted that the parrot had lived with Mrs Snooke for a 'full 19 years' and that it died on 12 December 1790. Betty's elegy was written as follows:

> Drop a tear, poor Poll is dead.
> Her friends surround her funeral bed.
> No more can hear her pretty prattle.
> For she is gone, no more to rattle.
> Her mistress has her coffin neatly made.
> With true affection and without parade;
> With weeping eye, she took her last Adieu and said
> 'Alas my dearest bird, I weep for you'.

We can only speculate as to why, according to Richard's fastidious diary, Poll was buried on 22 December, a full ten days after her death. Was it to wait for that little 'neatly made' coffin? Or to wait until the ground thawed? Perhaps it was on account of Mrs Snooke's difficulty in parting with her companion of 19 years. After Richard's death at the age of 72 in 1801 his widow, Betty, retained most of his possessions until her own death, 17 years later. She kept his books of clippings, his accounts and his diaries. The 1818 probate records of her belongings also listed, among many other things, a 'bird cage and brass trivet'; this was probably the very cage that had housed Polly decades earlier.[1]

The story of Richard Hall fits into a much broader story found here in *The Georgian Menagerie*, that of the taste for exotic birds and animals in London in the long eighteenth century.[2] In this period exotic birds and animals not only appeared in greater number and diversity alongside an increase in shipping and international trade, they could also be seen, bought, or sold in a variety of new spaces that emerged: in menageries, at animal merchants', and bird sellers'. This cultural taste was part of a trend for natural history that characterized eighteenth-century intellectual and polite society as part of the Enlightenment. This book examines the place of exotic birds and animals in the cultural landscape of eighteenth-century London. It traces the emergence of the trade in exotic birds and animals in London and explores the multitude of meanings they held and their significance to eighteenth-century Londoners.

Introduction

Some East Indies Ships arrived and brought some very strange birds and beasts, such as were never seen in England. A great number of persons from the city of London and other parts flock daily to see the rarities that they have brought over.

Tuesday's Journal, 24 July 1649

W HEN TUESDAY'S JOURNAL BROUGHT attention to the arrival of 'some very strange birds and beasts' in London in 1649, those Londoners that flocked to see them could not have imagined how events in the following 15 years or so would substantially alter life in their city. King Charles I had been executed only few months earlier, but the new Commonwealth would last little over a decade. The restoration of the monarchy in 1660 would be followed by the Plague of 1665 in which roughly one in five Londoners would die, and many of the living would flee the city. And then, in the very next year, the Great Fire of London would, in a matter of days, raze to the ground the homes of some 70,000 of the city's 80,000 inhabitants. Londoners would pick through the ashen rubble of their homes, with countless numbers dying in the following winter. Yet, the city would be rebuilt and grow rapidly in both size and status. By the end of the eighteenth century London boasted a population of around 900,000. It was the largest city in Europe, a metropolis into which poured the goods – and birds and animals – of the world.

The Georgian Menagerie is a history of the exotic animals and birds that increasingly became part of London life in the eighteenth century. They were sold in the premises of bird sellers and animal merchants, and placed on display in taverns or menageries. Foreign birds and beasts occupied not only the places where Londoners lived, worked, and visited but also their minds; in Georgian cultural life the ways in which these birds and animals were written or spoken about expressed a wide range of cultural attitudes or concerns. Letters, diaries, journals, legal records, newspapers, handbills, trade cards, poems, jokes, novels and natural history books all contain rich references to the menagerie that was Georgian London. Although privileged gentry and aristocracy – the fashionable and rich metropolitan elite, or 'Beau Monde' – play an important role, Londoners in this book are not only those of the well-born or well-to-do middling sort. Instead, *The Georgian Menagerie* is also the story of more ordinary Londoners; the bird catchers, the itinerant bird sellers, the canary fanciers of humble means, the bird thieves, the wigmakers, the workers in the turtle warehouse, the barbers, the servants, the maids and the keepers.

The focus of this book is London and the city's environs, though in this period exotic beasts and birds became increasingly present elsewhere too. Travelling menageries visited provincial towns, and exotic birds and animals adorned the landscape gardens of the aristocracy and gentry. The Georgian menagerie included some other cities such as Oxford, Bristol, Edinburgh and Dublin, and this book reflects London and its broader British context. Since the formal unification of England and Scotland in 1707, great efforts had been made to decipher, promulgate and contest the values and postures of 'Britishness' for a new United Kingdom. 'British' and 'Britons' are used here to designate shared interests and concerns, ideas and identities beyond London and the English throughout the British Isles. Eighteenth-century Londoners were drawn from all corners of Great Britain and Ireland, and even further afield. Sources in this book include diaries and journals of Scots, Irishmen, Americans and Frenchmen.

It is worth mentioning here that while our British story was unfolding around and in London, exotic animals and birds were also present in mainland Europe. The place and significance of birds and beasts in France, Spain, the United Provinces, and the German-speaking lands took on different characteristics and contours, however, shaped by their politics, languages and peoples.[1] When reading this book, then, it is important to bear in mind that as Londoners sold, bought or saw exotic birds and animals, so did their Parisian and Viennese counterparts.

The foreign birds and beasts brought to London during the eighteenth century were imported from Africa, the Americas, Asia and later from Australia. The birds and animals in this book were shipped to Britain with the intent that their presence there was desirable, meaningful and, perhaps most importantly, profitable. These animals were, broadly speaking, mammals, birds and reptiles. They were animals that could be readily supplied with their preferred food source or forced to eat another, and were resilient enough to, if not flourish, at least survive in Britain for significant periods of time. In the pages of this book are a menagerie of monkeys, parrots, zebra, camels, elephants, moose, lions, tigers, electric eels, crocodiles, pythons, kangaroos and llamas.

The adage that 'familiarity breeds contempt' is especially true in the case of the exotic and novel. Canaries, for example, were at first expensive imported luxuries, but within decades they were cheap enough that London's labourers and poorer tradesmen could reasonably hope to own one. Similarly, London spectators often expressed disdain or ennui for those menageries that had displayed a small range of familiar animals for too long. Proprietors assiduously advertised their animals as 'new', 'just arrived', 'the only of its species', or 'the largest of its kind' to try and appeal to an easily jaded urban audience.

Some advertised their stock with a certain measure of self-aggrandizement; in 1752 a handbill announced the exciting arrival of 'the grandest collection of wild beasts' at Charing Cross, including a crocodile and rhinoceros. Both these animals, it was claimed, were the only animals of their kind in the kingdom. This was the wild animal exhibition that Richard Hall later saw at the

Talbot Inn, pasting the handbill in his scrapbook. The novelty of these headliners was in turn underscored by the self-deprecatingly modest assertion that the other menagerie occupants were too numerous and far 'too tedious to mention'. Even the hyperbole of a showman can tell us that some exotic animals could become overly-familiar, which attests the surprising everydayness of exotic animals in eighteenth-century London.

Looking at these exotic animals makes it possible to understand more about the cultural preoccupations and social relations of Londoners during the long eighteenth century. The story of exotic animals in Britain is intertwined with the histories of trade, political satire, gender relations, humour and the changing urban geography of London. This book is about the meanings of exotic animals in Georgian culture, and the way these animals can be used to think historically about this period. The history of exotic animals in Georgian Britain is not very well known, so *The Georgian Menagerie* may be surprising in its accounts of erotic electric eels, 'The Queen's Ass', parrot suitors and turtle soup.

We begin with the story of the city's trade in exotic birds and animals.

PART I ❦ TRADE

IN MARCH 1704 DAVID Randal, a bird merchant who traded from his house in Channel-Row, Westminster, advertised his stock that included 'choice singing canary birds' and 'fine talking parrots' as well as a 'sweet monkey that whistles like a bird'. Randal promised to furnish these exotics at 'reasonable rates' and, trading at the beginning of the eighteenth century, was not the sole dealer in London. Indeed, his advertisement assured the public that 'a great many' traded in his name, but if persons of quality were seeking birds then they ought to come to him instead and 'be kindly dealt with'.[1] These bird sellers and merchants were trading in a city that was well supplied by the ships that unloaded wares from across the world on docks and wharves. The canaries, monkeys and parrots on sale attest the status of London as a hub; likewise, the sheer number of bird sellers and animal merchants in the city demonstrate a thirst for novelty and the exotic. Those with deep enough pockets to pay for birds and beasts had no shortage of vendors willing to oblige them and help part them from their cash. The following three chapters chart the development of the trade in exotic birds and animals in London, starting with the itinerant bird sellers and canary shops. The world of these dealers as well as that of the animal merchants is explored through the court cases, insurance policies and wills that leave traces of their lives. The third chapter considers the lives of these birds and animals once they were purchased or given as gifts and became the property of new owners. One such bird, a parrot called Miss Poll, became a singing sensation. Others lived lives away from the limelight, but are recorded in the diaries and letters of their owners or in the pages of the naturalists who visited the houses of the owners to draw foreign birds and beasts.

CHAPTER 1

'Buy a Fine Singing Bird'

In *The Cryes of London*, a series of seventeenth-century engravings of London tradespeople 'drawne after the life', a young man is pictured with his wares. He is selling canaries or songbirds and raising the cry of his itinerant trade: 'Buy a Fine Singing Bird'. In the bloom of youth and fashionably attired in buckled shoes and a tricorn hat, the birdman is the cheerful romantic face of the bird trade. The tradition of catching and rearing native songbirds such as nightingales, bullfinches and thrushes has a history in Britain, so by the seventeenth century a substantial trade and knowledge about bird catching and rearing had been established. This trade was practised by itinerants or tradesmen with small rented premises and was sustained by both the seasonal capture of birds and the demand for the novelty of birdsong in people's homes. However, life could be hard for bird sellers and their allied tradesmen – the cage makers and bird catchers – particularly for those in the lower ranks. The unhappy case of Thomas Roberts and John Girle from the mid-1750s is a brief glimpse into the lives of the itinerant and poorer birdmen of London.

John Girle was a young man of 25 years, born into a family of bird catchers who made a living trapping birds outside London between October and March. Bird catching was a trade that demanded a wide range of skills in luring, trapping and rearing. Further, a bird catcher needed the specialist vocabulary and know-how to identify species, common ailments and cures. It was from his parents that Girle learnt his profession and when he was orphaned as an adolescent it was with this trade that he made a living. Girle would sell birds on to bird sellers with their own premises.

This was how he encountered Thomas Roberts, a bird seller who traded from a humble rented cellar in Holborn. Roberts was described in court records as a 'cripple', pitied by witnesses for supporting his wife and two infant children on next to nothing. According to his wife, he was a cage maker and bird catcher *because* he was a 'cripple', not as an avocation. In reality, he was principally a cage maker who sold birds as a means to make a more substantial living. Roberts had acquired birds from Girle while trying to trap his own for years, but in 1755 Girle was unable to persuade Roberts to buy. Turned away empty-handed, and having tried to drown his troubles in drink, an angry Girle paid Roberts a visit, smashing a small cellar window with a stick and destroying a cage that had been hanging on display. With her children in her arms, Sarah Roberts had tried in vain to protect her husband's business. The sound of Girle 'knocking the cages' about carried from the cellar to the street outside and premises above. Witnesses gathered as the altercation reached its tragic climax. Roberts put himself between Girle's stick and his birds and lost an eye. The wound was deep, bloody and painful, and there was no money to pay a surgeon to attend the scene. His injuries were tended with rosewater only. Roberts struggled on with his trade for five months, partially blinded and in much pain. In the entire catching season he was unable to trap a single bird. The business was in peril. A surgeon was eventually called, but this only served to deepen the financial ruin Roberts's lingering death brought the family. Girle was tried for murder and convicted. The Court Ordinaries report of the last days of Girle's life portrayed him as a man resigned to his fate and willing to admit his wrongdoing. Yet another drunken altercation had become murder, the court decided, and so on 17 July 1756 John Girle was sent to Tyburn and hanged.[1]

For men like Girle and Roberts, bird catching and selling was a difficult and barely practicable trade. Roaming bird catchers needed bulk buyers if they were unable to shift stock individually. Meanwhile, cage makers eager to supplement their earnings from wood and wire were looking to sell birds on the side, preferably catching their own birds for free. An early eighteenth-century guide to the trades of London reported that cage makers expected to make 12–18 shillings a week. A haughty note states that although the trade required neither 'special strength nor skill', it was in demand, at least. The 'bird fanciers in and around London' were 'so numerous a tribe' that the services of a cage maker were always required.[2] Rattan and cheaply made wooden cages could be purchased inexpensively. Even a smart small mahogany cage might not fetch more than a shilling or so. Many simple cages were either small wooden boxes with wire fronts, or larger, mostly wire-bodied domes with sparse wooden bases and supports. Birds were, however, more expensive than the cages that housed them. Even native songbirds fetched at least a shilling, and imported canaries something in the order of 10–15 shillings for a choice bird. A pair of canaries suitable for breeding might make the handsome sum of 20–30 shillings. By the last decade of the eighteenth century, canaries could be bought for as little as a shilling. It became clear to cage makers that the money was in the birds and not their housing, so many sought to keep a finger in both pies. As such, most seventeenth-century and early eighteenth-century cage makers were also birdmen. Around 1700, of all the songbirds kept by Londoners, it was especially canaries that were more costly than the cages that held them.

Canaries were a staple of the bird trade by the close of the eighteenth-century. They first entered the London market in the early 1700s from the Spanish Canaries, imported with wines, although most birds in our period came from German-speaking lands and bird-sellers called canaries the 'German birds'. They had been selected in captivity by breeders in Central Europe for their singing voice, with several varieties commonly available for purchase: white, yellow and brown.

In France, those German bird sellers competed with the Oiseleurs, a guild of bird trappers and sellers. In response, the Oiseleurs rallied to defend their privileged monopoly in the bird trade. In England, there was no such guild structure for bird sellers. One of the early prominent bird sellers in London was known as J.C. Meyer, 'the German', with premises on St James's Street. Established by 1700, Meyer's shop sold all manner of necessities and fripperies to tempt the committed or casual canary fancier, from many assorted bird seeds to handcrafted cages after his own ingenious design. Meyer had developed line-and-pulley contraptions to suspend and lower his ornamental cages; they were furnished with the very finest of novel accoutrements, including crystal glasses to hold seeds. With an eye on the competition, Meyer strenuously insisted in his advertisements that his goods were of vastly superior quality and design to those hawked on the streets by itinerants. Initially, Meyer's rivals mostly took the form of birdmen who sold in the streets on an ad hoc basis, or irregularly traded imported 'packets' of canaries from taverns or coffee houses. Dedicated bird sellers in established premises began to appear more regularly in the newspapers and periodicals of London a decade or so later.[3] Bird merchants such as David Randal at Channel Row, Westminster, also complained about the competition. Randal sold parrots and monkeys in addition to his canaries, and his advertisement of 1704 shows that he was clearly not the only merchant to enter the trade at the turn of the eighteenth century.

One such birdman was Thomas Ward, who began trading from the Bell and Bird Cage on Wood Street. Ward drew on his 20-year experience as a bird catcher, his original trade. His business was a profitable one, eventually allowing him also to operate from the Bird Cage on Stamford Hill. From his premises, Ward sold a self-penned pamphlet on the appropriate selection, care and breeding of canaries and native songbirds. There were three editions between 1727 and 1740.[4] The compendium was an authoritative entrée into the world of bird keeping and established Ward as an authority. There was a certain art to choosing a 'fine singing bird', with several pitfalls to avoid. The undiscerning buyer might be tricked into buying a sick bird. Unscrupulous bird sellers would pass their hand over the birdcages in an ostentatious gesture that drew the untrained eye to the birds. These scoundrels were agitating the birds to make even the most sickly appear sprightly and bright-eyed. Worse still, some might try to pass off dowdier female canaries that struggled to hold a decent tune as more expensive male songsters. Except for breeding purposes, a male was strongly preferred, since a cock's 'singing' contrasted with a hen's regrettable 'jabbering'. One could sex a canary by lightly blowing feathers away in the ventral region to reveal the sex organs. Removing a bird from a large store cage away from other birds was likewise the best way to determine if a bird was indeed sprightly or sickly. Other handbooks for would-be canary fanciers advised that to pick the healthiest canary, 'the greatest matter' was to 'observe his dunging'. Dung that was 'thin and long like water' was a sure sign of approaching death, in contrast with the hard, white, rapidly drying dung of a healthy bird. One ought to listen carefully to the sounds a prospective bird uttered. Ward thus provided his market with a solid grounding in the choosing of suitable birds: by implication, the ones he sold.

Once a fine singing bird had been bought, he needed to be trained to sing. From 1717, the anonymous pamphlet *The Bird Fancier's Delight: Choice Observations and Directions Concerning the Teaching of All Sorts of Singing Birds* was available for guidance. Using a woodwind flute called the bird flageolet would teach a bird a simple tune. A music-loving bird owner would learn a piece for the flageolet and repeat it daily, or even several times daily, in the hope that the feathered listener would pick up the tune. Fanciers with more money to spend might buy a mechanical French bird organ known as a *serinete*. This small barrel organ could play a larger repertoire of songs without fatiguing a human musician. Though, for the would-be songster, it involved more regular and longer music tutorials. Coercion could take a more sinister turn, as it was found that the best singers were those that underwent instruction and isolation in the dark. John Rambling's *New and Complete Bird Fancier* instructed despairing owners in the process of 'stopping'. Once over a year old a canary or other songbird could be enclosed in a small cage, which would then be covered in a dark cloth. In this warm, dark and musty cage the bird would moult. The cage was only rarely cleaned, 'by reason the hotness of the dung forces them to moult'. After three months the bird would have thoroughly moulted, and then the cage was slowly exposed to the light again. The newly feathered bird, relishing the natural light, would sing 'still more and more'. The lively songster could be used thereafter to train younger birds or act as a stool pigeon, used by bird catchers to lure in wild birds. Naturally, books printed for bird fanciers also recommended suitable diets for birds. Typically these included millet, hempseed and canary seed. Canary seed, though initially imported, was by the early eighteenth century established as a domestically grown agricultural product. Much of the canary seed in England was grown in north Kent, on the marshy Isle of Thanet, where in addition to being a cash crop, canary seed was thought to take some of the 'rankness' out of the salty soil.

Birdmen were not only a source of know-how on selecting and rearing birds; they were also an invaluable resource for naturalists interested in the origins, behaviours and natural history of foreign birds. Eleazar Albin, a naturalist and illustrator, called upon the bird sellers to furnish him with tidbits of information and grant him opportunities to draw the birds from life. In the footnotes of his *A Natural History of Birds* (1731–8) he credited birdmen as the sources of birds, especially the birdman at the Tiger Tavern on Tower Hill. This particular tavern must have had a rather impressive range of birds to be seen and sold – well beyond the mundane canaries or chaffinches – since Albin recorded and drew two cassowaries and an Indian vulture there. The cassowaries, he noted, could be seen greedily eating bread, fruit and meat with their tongues. Along with the vulture, the cassowaries had been brought over from the Dutch East Indies, and their keepers seemed to know something of their habits. Other birdmen could offer less specific knowledge about the origins of the birds they sold. George Edwards, librarian to the Royal College of Physicians, had himself a veritable menagerie of exotic birds and animals that included parrots, finches, a lemur, a squirrel, a monkey and a pair of tortoises. Like Albin, he used taverns and bird dealers as an important source of knowledge and drawings for his

Natural History of Uncommon Birds (1751). Listing in his footnotes the sources of his birds, Edwards mentioned several birdmen operating in London, usually giving the location of their premises, such as Crooked Lane or the Strand. Edwards would sometimes note that the seller 'could not inform' him of much of value about the bird; instead Edwards had to take an educated guess about the bird's origins. His green parrot was believed to be from the West Indies, since that was where most were imported from. The green parrot was 'talkative in a language unknown' to Edwards; a pity since sometimes the language a parrot had been trained to speak, typically French or Spanish, might have hinted at the origin or journey of the bird. His parrot really must have been prattling in quite an unusual tongue, since Edwards was a well-travelled man, having journeyed through Europe as a young man.

The trade of the bird catchers and the world of bird fancying was one ripe for satire. In his *A Complete and Humorous Account of All the Remarkable Clubs and Societies in London and Westminster*, John Ward conjured up for his readers' pleasure the world of the city's birdmen. His description of the 'Bird-Fanciers Club', though designed to lampoon, reveals something of the social attitudes towards the bird trade and bird fanciers. This club gathered in a small alehouse on Rosemary Lane, no bigger than a cage, small but well suited for those whose affection for the 'feathered kind rendered them fitter company for a jackdaw or magpie' than their own kind. Around the tables huddled a miscellany of canary merchants, journeyman flute makers and cage makers. The 'lousy bird catchers' were dressed in rags, their clothes deemed suitable neither for a dung heap nor for a rag merchant's warehouse. The itinerant bird sellers were little better: 'Newgate scoundrels' – common criminals – who roamed the streets like 'crying singing birds', making it their business to 'cheat barren wives and old maids'. The bird trades were complemented by the inclusion in the company of a cage maker who was eager to engage all in discussion of the latest in 'cage architecture'. The Bird-Fanciers Club was completed by a tavern keeper. The publican's love of birds had turned his alehouse into an aviary, more to his pleasure than his patrons', so that the birds might 'shite flying on people's heads' or even 'muddy their drinks' with hempseed. This motley assembly of fanciers was characterized by low-status, criminality and disrepute, or by eccentric obsession. Ward's comedic scene reached its denouement at the club's annual feast. A 'live-bird pie' was brought to the table and cut open, with the assembled birdmen scrambling like cats in 'hair brained pursuit' of their feathered quarry. Once caught, these birds were given by the birdmen as tokens of affection to their wives and daughters.[5]

Ward's satire of the bird fanciers and birdmen was grounded in perceptions of unscrupulous bird catchers, disreputable bird sellers and unkempt obsession. This was no doubt exaggerated but, like all satire, contained some seed of truth. The grim case of John Girle's murder of Thomas Roberts may have been exceptional, but songbird and canary theft appears with quite some frequency in London legal cases throughout the eighteenth century. These thefts were typically motivated by the resale value of canaries or the avarice of fanciers. William Frewen, a plasterer of St Giles, took the man who had moved his worldly goods to new accommodations – and in the process made off with two birdcages and five canaries worth ten shillings – to court. The opportunist thief, Edward Riley, had taken the birds to a tavern in Moorfields and attempted to sell them on. Leaving one of the cages and three of the birds in the tavern, Riley had tried to sell a pair of canaries on to a bird fancier, a man who acted on his suspicions and reported Riley. For his pains Riley was found guilty of feloniously stealing the canaries and whipped. In a similar case of domestic canary theft, the hapless criminal was caught at seven in the morning going over a wall, with a canary 'found in his breeches'. This canary crook, hoping to make an easy shilling, got more than he bargained for and was sentenced to seven years' deportation.

Perhaps the most elaborate and embittered case of canary theft was that of tavern keeper John Smith's canaries on 7 September 1743. It was heard at the Old Bailey in December 1744. Four canaries in two cages were allegedly stolen from Smith's home by his own servant, John Marshall, a carpenter by trade, lodging at the tavern and carrying out odd jobs for his board. One such job was constructing a wooden partition for an aviary. Clearly, John Smith was the sort of tavern keeper who was lampooned for turning his alehouse into an aviary. Before the relationship between the two men soured, they had shared an appreciation for canaries; Smith having gifted him some canaries, which Marshall kept at his mother's house. After the theft, Smith acquired a search warrant and went to the house of his former servant's mother. In her home there were five cages with canary birds inside – only one had belonged to Smith. In court, a picture of Marshall as a canary thief was painted. Especially since, so Smith claimed, Marshall had 'pulled the feathers off' the others. He had found one bird – but surely the others were his too. In Marshall's defence, witnesses testified to a rather different account. Apparently the two men had regularly exchanged birds as breeders and when sick, these birds had lodged at Marshall's mother's house. Marshall's father – living in a different abode – was also a breeder of canaries. As such, these men formed a small circle of bird fanciers and sellers, exchanging birds with one another. The plot thickened further once John Smith was found to have forced another servant, Sarah Wrench, to testify against Marshall. The tavern had been having financial problems; Smith complained that he was unable to pay the distiller for liquor. Perhaps Marshall had had his hands in the cash box – Smith had threatened and assaulted Marshall, accusing him of theft from the tavern's till. Ultimately Marshall was acquitted of the charges of theft. Smith had come off worse. It seems he was maliciously seeking the return of a gift, perhaps even coveting the birds that belonged by all rights to Marshall. The waters of this tit-for-tat case were so muddied by scattered hempseed it was difficult to ascertain guilt. These canary court cases are a small glimpse into the circles of birdmen in Georgian London; at its best a world of mutual interests and cooperation, at its worst a murky world of competition and conniving canary charlatans.[6]

The prominence of coffee houses and taverns in the history of London's early bird trade is not an anomaly; these sites were important spaces for trade and social life. The cultural taste for coffee had emerged in the 1650s and 1660s, and coffee houses became an important part of what is called the 'public sphere'. Coffee houses were not merely important hosts for auctions, insurance agents and merchants; they had an important role in the history of the print press and sharing of political opinion. After the Civil Wars of the 1640s and 1650s coffee houses were important spaces for political debate and dissent, as well as cultivating conversation. Both coffee houses and taverns hosted the meetings of a diverse range of societies; ranging from gentlemanly political clubs and learned societies, to bird fanciers. This 'public sphere' was also exhibitory, for in the spaces of the coffee shop and tavern exotic birds and animals were both displayed and sold.

During the eighteenth century the importance of the coffee house and tavern as retail spaces diminished, in part due to a change in consumer habits. Economic histories of the long eighteenth century suggest that in the period 1675 to 1725 there was a dramatic increase in the introduction of material goods into domestic spaces in Britain, especially London. This was in the same decades that birdmen like Meyer and Ward were establishing their premises, offering canaries and cages to customers. Goods that had been hitherto rare in a domestic context – far out of the reaches of the pockets and purses of the average Londoner – became increasingly commonplace. The ownership of clocks, for example, tripled in the early eighteenth century. And, although rare in 1690, by 1725 utensils and china could be found in use in a much larger number of households. New goods are an indication of new shopping habits and cultural tastes. An increase in the ownership of ordinary goods was accompanied by the consumption of products increasingly imported at relatively modest prices. In a world of material goods the practice of shopping and the physical space of shops altered in the eighteenth century as shopping became a staged pleasure; an art of selecting between varieties of manufactured goods on display. In the early eighteenth century, shops might range from fairly modest wooden stalls or counters to the fronts of houses converted to trade purposes by the fitting of glass windows. However, by the late eighteenth century, the purpose-built shop and showroom emerged as a retail space replete with decorative mouldings, cabinets, cornices and pillars. Shopkeepers began to develop specific strategies for selling their goods, such as glaze-panelled windows with crossbars to frame the window and draw the attention of shoppers onto particular goods, along with racks, display boards and cases. In London and larger towns, shops and shopping became part of a wider public culture of consumption. The selection of goods whilst promenading around shops was a manner of performance on the street or in the arcade that worked alongside the theatre and assembly rooms as spaces to see and be seen. The emergence of specialist bird shops, alongside itinerant sellers and inns, fits into this story of economic expansion and a change in shopping habits. Canary birds – although fine singing birds – stopped being foreign novelties, as bird sellers were able to offer more exotic stock.

Away from the city and in the country, by the mid-century, farmers too were experimenting with raising exotic birds and breeding them as a means to pay rents. William Ellis, a farmer of Hempstead in Hertfordshire, described in his *Country Housewife's Family Companion* (1750) how to raise peacocks, peahens, guinea hens and gamebirds. The peacock with its 'pompous spreading gaiety of fine-coloured feathers' was a fine addition to a farm as it was thought to eat snakes and lizards and would roost on the ridges of houses, in trees and in barns, enduring even the frost outside. Peacocks and peahens could not be trusted near an orchard, however, as they would pilfer and damage the fruit. Peacocks could be quite an earner for farmers, with a male fetching half a crown. The investment in a few peafowl purchased from a neighbour or London dealer could pay dividends. Not all these birds were destined for the table; Ellis supposed that some could live to 30 or 40 years. Ellis's writing on rural economy was based on his experiences in Hertfordshire and Bedfordshire, indicating that mid-century enterprising farmers in the southern counties were introducing exotic fowl into the rural landscape and dabbling in the bird trade.[7]

The 1750s and 1760s were decades of war and economic expansion for Britain, a period that changed the character of bird merchants in London. British territorial gains and engagement in the Seven Years War (1756–63) seriously diminished French authority overseas. Britain acquired lands in North America and India at the expense of the French, and the East India Company grew substantially in these years. The rapid increase in shipping between Asia and Britain was immense. In the instance of tea, for example, in the early 1700s around 200,000 lb per annum was carried by the British East Indies Company, but by the late 1750s this had swollen to 3,000,000 lb annually. Similarly, the volume of shipping entering the Port of London increased fourfold during the eighteenth century. Ships returning from overseas territories and foreign ports brought with them exotic birds and animals in increased variety and abundance. Advertisements in newspapers and periodicals show that from the 1750s and 1760s bird and animal dealers were selling a broader range of species than the bird sellers of the early eighteenth century, many originating from newly acquired lands. A varied urban jungle emerged, consisting of specialist bird sellers offering increasingly exotic avian species to bird fanciers, menagerists dealing principally in four-footed exotica, as well as those in between. Many, like Edmond's Menagerie on Piccadilly, advertised a greater range of birds in the 1760s than had been available earlier in the eighteenth century. There were more parrots, fewer canaries and small novelty mammals such as American chipmunks. The presence of songbirds like the dazzlingly ruby-red cardinal and animals like box turtles and chipmunks demonstrates an increased availability of North American species. This is a reflection of both increased trade with the North American colonies as well as territorial expansion on that continent after the Seven Years War. Other merchants such as the City Menagerie had a considerably larger repertoire of animals for sale in addition to the usual feathered fare: monkeys, tigers, opossums and camels were all advertised as 'to be seen or sold'. The animal merchant dealing from the very aptly named Noah's

Ark offered a wolf, crocodile, several camels and a huge variety of parrots and other caged birds.[8]

In the mid eighteenth-century the sale of exotica moved well beyond fine singing birds and, although the itinerant bird sellers, canary fanciers and birdmen with small premises continued to ply their trade, they were joined by a new sort of bird and animal retailer: the menagerist or animal merchant. One of these was Joshua Brookes, a man from a family whose position had been secured by their skills in bird handling and breeding.[9] From a family of Dutch bakers with the surname Brük, one such Brük was a falconer to William of Orange, who found that his fortunes improved when he crossed the Channel along with his patron. Brük, later known as Brook, received a knighthood and married Lady Jane Marmon; their grandson Samuel Brookes, eager to move the family name away from humble loaves of bread to a landed origin, added the Latinate 'es' to suggest a more aristocratic origin. Samuel Brookes was the father of Joshua Brookes. Brookes's menagerie represented the 'high end' of the trade. In his earlier days Brookes was principally a birdman; from the 1750s and early 1760s he offered birds from Bengal and America for sale, trading from his 'Original Menagerie' at Gray's Inn Road, Holborn; indeed Brookes and Thomas Roberts – the murdered bird seller, trading from a cellar-shop – were neighbours and contemporaries. However, Brookes's livelihood was a stark contrast to Roberts's poverty. From around 1765 Brookes extended his premises to include a menagerie on the New Road at Tottenham Court, and these premises were more opulent. A turreted elegant building with large windows, standing in well-tended and planted grounds, his premises were advertised as well kept and ventilated rooms where exotic animals could be bought or exchanged. Although a catalogue available to customers listed species as diverse as antelope, lions, monkeys and porcupines, it was principally in birds that Brookes made his trade in the early years. A handbill printed to advertise his business in 1775 lists some 160 different avian species alone, ranging from cassowaries to cockatoos and finches to flamingos. Brookes promoted himself as a prominent and capable 'zoologist' able to convey his birds from and to 'any part of the world'. Brookes's professional identity was diffuse, with him sometimes advertising himself grandiosely as a 'zoologist' and other times rather more modestly as a 'bird merchant'. It is clear, though, that the basis of his early businesses was grounded firmly in the bird trade; trading exotic animals was a lucrative extension of his core activities. A small elegant trade card for his Holborn business states simply 'Joshua Brookes, Bird Merchant. Buys, Sells & Exchanges all Kinds of Curious English and Foreign Fowls, Birds, &c. at reasonable rates'. His reasonable rates aside, at first exotic animals were very much an 'etcetera' for Brookes.[10]

In the 1780s and 1790s West London was peppered with an array of retailers catering to the fashionable tastes of the metropolitan elite who resided in grand squares such as St James's or the great aristocratic houses like Devonshire or Hertford House. In this milieu, Twining's tea merchants and Fortnum and Mason stocked the tables of the wealthy with the edible commodities of empire; Drury's provided the silverware. Josiah Wedgwood's showroom on St James's Street and the print shop of Rudolph Ackerman on the Strand were spaces for conspicuous and luxurious consumption, pleasurable socializing and leisurely browsing. Bird merchants were able to tap into this market by setting up shop in the West End too. By the 1780s the simply named Bird Shop was trading from the Haymarket, Piccadilly. Around 1790, Wyatt of 192 Oxford Street, near Portman Square, was both a cage maker and a dealer in 'foreign' birds, with a trade card featuring a rather noble eagle perched astride an elegant cage. Menagerists dealing in exotic animals also operated in this part of London: Mr Kendrick's Menagerie at 42 Piccadilly and Gilbert Pidcock at Pidcock's Grand Menagerie, Exeter Exchange (or 'Change). The late eighteenth-century bird and animal trade was principally located in two areas: in West London along the Strand, Piccadilly and Oxford Street; and further away at the boundary of the City of London, in Holborn. Indeed, in Holborn in addition to Brookes's Menagerie one could also find Gough's Menagerie selling birds and beasts, as well as Grainger's. In Regency London the grandest, or perhaps most self-aggrandizing, bird seller in London was Grainger, 'Purveyor of Birds to His Royal Highness, George Prince of Wales'. From his premises in Gray's Inn Lane, Holborn, with the 'Prince's arms over the door', Grainger sold a variety of exotic birds. Clearly peeved that lesser bird merchants were using and sullying his name, Grainger's trade card sought to set the public straight: 'many impositions have been made on the public by persons making use of my name'. In the competitive world of bird selling, the right name went a long way. Grainger's establishment was clearly no humble canary shop, and likewise James Pilton's Manufactory was not a mere cage maker. A short carriage-ride from West London, in leafy Chelsea, was Pilton's manufactory, which from the 1790s offered a comprehensive service to wealthy clients. Pilton, principally a designer and producer of decorative ironwork, provided his services to the gentry and aristocracy. The menageries and aviaries that Pilton provided were 'judiciously arranged and stocked' with feathered inhabitants. The trade cards and promotional material for the manufactory depict the premises as a series of gardens in the grounds of a house, a paradise in which peacocks and pheasants roam amongst elegant ironwork structures. Like Grainger, Pilton's handbills drew attention to the esteemed patronage of his business by 'their Majesties and the Royal Family'. Certainly, Pilton represented the very pinnacle of his trade.

The lives of the early bird sellers are not historically well documented. It is a history to be gleaned from advertisements placed in newspapers, proceedings of criminal cases and the footnotes of natural history books. The Georgian animal merchants and menagerists are, however, a different case, since several made wills and registered insurance policies. The judicious household management and prudent business acumen of these tradesmen, as well as a few women too, allow a historian to peek into the everyday material lives of those involved in the animal trade. The following chapter probes into the private lives of those who sold and displayed exotic animals and birds in the last half of the eighteenth century.

CHAPTER 2

'To Be Seen or Sold'

ARLY IN THE MORNING, so a young Irish gentleman studying law in London wrote home to tell his father,

> I was walking down Piccadilly, when I met a porter carrying a live kangaroo, which he was conveying from Mr Pidcock's at the Exeter 'Change, to a person who had purchased it. The animal was fastened to his knot by the feet, and his head lay dangling over, very near the left ear of the fellow who was carrying him; this it seemed was a temptation not to be resisted by the kangaroo, who, after smelling at the man's ear for a long time, gave it a terrible bite, and nearly clipped it off.

But this was not the end of this astonishing anecdote; our student regaled his father with a detailed description of the ensuing street fight. Apparently in a rage,

> the porter threw the kangaroo with all his force upon the pavement, which completely stunned the poor kangaroo upon which a drayman that was riding past ran to the porter and immediately jumped to defend the poor animal. A crowd speedily collected, a ring was formed, and the drayman after several severe rounds gave the porter a hearty drubbing. Then he took the kangaroo into his cart and as the animal had the address of the person to whom he was going fastened to the string that confined his feet, he drove off with it to deliver it followed behind by the porter whose head and face from the united

exertions of the kangaroo and his protector presented a hideous spectacle.[1]

This spectacle, notable as it was for a 'hearty drubbing' and a kangaroo with an address tag on its feet, was one of many close encounters with exotic animals in Georgian London. The kangaroo had been purchased from Pidcock's, a famed menagerie on the Strand; further along Piccadilly were the premises of other dealers, including Brookes. The commercial world of these dealers is not well known; they are absent even in histories of eighteenth-century commerce and consumption. To contemporaries this would have been a surprising omission, as the animal trade added conspicuous colour and novelty to the Georgian urban jungle. The names of the period's best dealers proliferated in print advertisements and were familiar to all but the least *au courant* Londoner. Even foreigners, clasping their printed guides to London, were directed to the dealers' premises to see, if not buy, their exotic spectacles. Where did one go to buy a kangaroo or a parrot? Who were the animal merchants?

The lives of a few of London's more prominent animal merchants and bird dealers can be traced in those documents that were intended to secure property – wills and insurance policies. In the eighteenth century, families of the 'middling sort', reliant on trade as an income, were frighteningly vulnerable to the turning tides of the world of commerce or the death of a husband and father. A hitherto respectable family might be catapulted to the bottom ranks, a fate befitting the family of a man who sat at gaming tables, but not that of the industrious bourgeois. Georgian Britain's burgeoning merchants, clergy, lawyers and prosperous shopkeepers or artisans were not of independent means as the gentry or aristocracy were. Fearful that their assiduously well-built and well-appointed houses of cards might easily topple, the middling sort did not have to look far to see the ominous gloomy clouds of foreboding. The mass unfortunates were right on their doorstep; some 70 to 75 per cent of the urban population alone laboured in service or was employed only seasonally and intermittently. They worked in the houses of others, laboured in the businesses or manufactories of others and, moreover, had slim hopes of improving their position. Worse off still were the poor or destitute who turned to poor relief or the workhouse. As a result, the early eighteenth century especially saw innovations in insurance and annuity societies, selling assurance to the middling sort. Death was at least one vicissitude of life that might be weathered reasonably by one's family if one made sufficient arrangements. So too, in a world illuminated by candles and oil lamps, was fire common enough to compel those property owners who could insure against loss to do so. Happily for the historian, the anxieties that weighed heavily on the shoulders of the propertied prompted them to leave records offering insight into their lives.

In particular the names Brookes, Pidcock, Polito and Cross were prominent in London as families involved in the animal and bird trade. Little is known about the other smaller, less prosperous and prominent dealers. From the 1750s the Brookes family emerged, as we saw, principally as bird traders; with their first menagerie based on Gray's Inn Road. Later, they extended their premises to other locations in London and began importing a wider variety of stock. The Brookes name was well known in London, and Joshua Brookes counted his clients among the aristocracy and the city's professional men, including the anatomist and surgeon John Hunter. Brookes could count on a network of overseas suppliers to secure stock; as such he was master of a business that was made on the high seas, good credit and a robust reputation. Without the promise of the costs of capturing, purchasing, feeding and transporting animals or birds being made good, Brookes would not have been able to source novel stock in sufficient quantities. It is clear that Brookes diversified too; from 1772 the 'King's Botanist' William Young began to send large quantities of plants and seeds to Brookes from the American colonies. Actually, Young was officially the Queen's botanist – he received an income from Queen Charlotte, a lover of botany, at her gardens at Kew. The largest consignment of plants Young shipped to Brookes included an astonishing several thousand flowering shrubs, trees, bog plants, ferns, seeds, acorns and cones for planting in gardens or preserving with stoves. The plants included orchids and the Venus flytrap. The flytrap was a sensational coup of botanizing and had been shipped alive for the first time by Young in 1762. This large botanical bounty was split into two different types of boxes from which customers could choose. Boxes with 44 varieties of plants along with a catalogue of their contents were sold for £4 4s., and boxes with around 90 varieties of seeds for £3 3s.[2] Brookes clearly knew the tastes of his wealthy and fashionable clients; the landed were increasingly reshaping their estates into landscaped gardens. Joshua Brookes, zoologist, as he fashioned himself, was just the man to help stock and plant them. Annual supplies of plants, and later birds and animals too, from the 'King's Botanist' gave Brookes an excellent market edge. Brookes's botanical and animal consignments ceased with the outbreak of the American War of Independence (1775–83) and he did not advertise these again until 1786. Brookes's business had to weather the storm of the revolutionary years, as conflict on land and the high seas disrupted trade routes. The cessation, at least of legal trade, between the rebellious 13 colonies, and later the early republic, by necessity compelled Brookes to source from elsewhere. In this decade, far fewer North American species are advertised for sale, and there is an increase in imports from Asia and Africa.

Expansion into exotic flora was accompanied elsewhere by the growth of the business. Brookes opened another menagerie at the Haymarket near Piccadilly, an excellent strategic site to take advantage of West London's elite residences. Joshua Brookes did not operate the menagerie at the Haymarket, however; he went into business with a new partner, Mary Cross. Brookes had been in business with her late husband John Cross, who had also been a proprietor of a menagerie in St James's until his death in 1776.[3] In his will Cross had left the bulk of his money to his wife, and smaller sums to his brother's widow and her sons. The Cross family would appear again as a prominent menagerie family later in the Georgian period. For now, though, their financial fate was intertwined with Brookes's. Cross's generosity extended to Joshua too, leaving him a gold watch. It was a token of friendship returned in Brookes's payment of Mary

Cross's fire insurance policy in 1777, a year after his friend's death. In the policy Brookes named Mary Cross as a 'dealer in live fowls and birds' and insured her (his) property for £300.[4] This property, 'two houses laid into one', was a substantial one, indicative of the money to be made in the trade. Like Mary Cross, many widows of the middling sort continued in their husbands' trade, especially if they had not been left in a secure condition. In supporting his former partner's widow Brookes was extending largesse to an extended kin-by-trade, though no doubt also hoping to make a handsome sum in the long run. Mary Cross was probably intimately involved in the operation of her business, by necessity if not will. Animal merchants, if they could afford to do so, employed servants but further down the pecking order, in the small dimly lit shops of the wireworkers and birdmen, servants would have been something of an unaffordable affectation. Women of the middling sort played an important role in the family economy and management. The degree to which this was 'hands on' was determined by the scale and status of the family. Joshua Brookes assiduously grew his business into the most prominent of its kind in London. Outliving his first business partner by some 25 years, Joshua Brookes left his second wife, Elizabeth, in a much less tenuous position than Mary Cross had been.

The will of Joshua Brookes, proprietor of a menagerie since the 1760s, is replete with annuities and estate stipulations – the hallmarks of the prudent middling sort striving to secure respectability and stave off a disastrous slip into poverty or destitution. Joshua Brookes was the owner of a small number of freehold and leasehold properties and the estates which, alongside annuities, provided financially for his family. Several decades of trade in exotic birds and animals allowed Brookes to leave his family in good stead. In 1803, on his death, his wife Elizabeth Brookes received £200 in a lump sum with a £50 annuity, supplemented further with an annual income from estates of £150. With a personal annual income of £200, Elizabeth Brookes would be comfortable, though by no means rich.[5] In the late Georgian period, a tradesman might need £40 or £50 a year to sustain a family but this would not be considered a genteel household. Likewise, a household servant in addition to lodgings and food might earn anything between £4 and £8 a year, with male servants earning more than females. With a secure income of £200 Mrs Brookes with some justification might have considered herself one of England's 'polite and commercial' middling sort – money enough for a few servants, for tea and new clothes and to engage in the sociability requisite of her sort. When the Brookes's menagerie on the Tottenham Court Road was finally sold in 1813 its contents were listed alongside the family's goods, including 'good pieces of fine mahogany furniture' commensurate with elegant living. We should not, though, imagine Mrs Brookes as an opulent widow; after all, the gentry and aristocracy who bought her late husband's animals and birds counted their annual incomes in thousands or even tens of thousands, not hundreds.

Joshua Brookes's will provided for the financial independence and respectability of not only of his wife, but also his daughters Arabella and Ann. Brookes had a substantial family he needed to provide for, with children from his first and second marriages. Both his unmarried daughters were in receipt of £50 annuities until they chose to marry, upon which they would receive £200 in lieu, presumably to constitute a modestly attractive dowry. His daughter Sara, married to Phillip Castang who ran Brookes's Menagerie on the Strand, received a lump sum of £200. The Castang family remained in the animal trade for generations; in the mid-1950s Phillip Castang Ltd operated from Hampstead as a seed merchant and supplier of aquarium equipment. In the office of Phillip Castang Ltd hung the trade cards and prints of Brookes's Menagerie. [6]

Brookes's sons George, William and Thomas were without an annuity but received £200 as a single payment, and his son Paul received a much smaller sum of £20. Clearly Joshua Brookes sought to protect his wife and daughters as best he could. His sons would have to continue in trade, relying on their wits and vocation. Joshua Brookes's son Joshua, who had studied anatomy and surgery, was given a leasehold estate on Blenheim Street, land that would become the site of his anatomy school and museum.

Shortly after 1803, with his £20 inheritance, Paul Brookes set out to establish himself. He did this by travelling the world to acquire birds and animals for his own business. This was rather different from the genteel Grand Tour through France, Italy and the Low Countries that the sons of the nobility and gentry who bought his late father's birds and animals undertook. Instead, Paul Brookes went first on a voyage to Africa and Asia, and later on a second journey to South America. Grand Tourists returned with a smattering of Italian, a few paintings and a sculpture or two. Brookes returned from his travels with 'a choice collection of curious quadrupeds and birds'. Reviving his father's old London premises on the Tottenham Court Road, near Fitzroy Square, Brookes printed his own trade card. This card, like his father's, pictured the same elegant menagerie premises, a fashionably crenulated Gothic-revival house. It was his 'great honour to inform the nobility and gentry' that he had returned from a voyage of several years to 'various parts of the globe'. This voyage had been for the purpose of 'collecting' and establishing a 'correspondence' which would allow him to obtain a 'regular supply of the most rare and interesting animals'. Like his father, he was reliant on a network of agents, sea captains and acquaintances to acquire his stock. In posturing as a well-travelled and savvy connoisseur of the exotic and rare, Brookes catered well for the tastes of elite Regency London.[7]

A male relative of Paul, perhaps an uncle, John Brookes, kept the family's menagerie on Piccadilly at the same time as Paul established his premises. John's business dealings reveal him to be something of a shrewd, slightly disreputable skinflint.

A snippet of John Brookes's private life is known through a celebrity scandal that reached the pages of the newspapers and periodicals in 1816. Living for six years 'as husband and wife, but not married' with a woman named Sarah Tookey, he was not leading quite the life of a prominent London merchant from a solid middling-sort family that one might expect. Attending to business affairs in Scotland, Brookes left his lover in the trust of some friends living at Westmorland Place in Pimlico. At that time Pimlico was a marshy district on the fringes of Buckingham House, characterized by taverns, market gardens, osier beds for cultivating willow trees for basket-making, and pleasure gardens replete with romantic arbours and alcoves. Not such a respectable Westminster address. He made sure, so the *Annual Register* informed its no doubt salivating readers, to make 'suitable provision for her maintenance' in his absence. However, whilst John was away, Sarah 'became acquainted' with another man, a Mr Thompson. Her new beau persuaded her to leave Westmorland Place two days before her jilted lover returned to London. What then transpired seems to belong to the well-thumbed pages of a tragic romance novel. A 'much grieved' Brookes searched London for Sarah's brother, a hairdresser in Ball Alley, George Yard, huddled in the City of London's financial district. Brookes left word that he wanted to speak with Thompson, and upon meeting him a few hours later a quarrel ensued. Physically smaller, Thompson had prepared a pistol knowing that he could not hope to overwhelm Brookes. Brookes armed himself with a fire poker in self-defence and gave Thompson a 'violent blow' to the head. The altercation continued, and the screams of women attracted more witnesses to the scene. At the very moment that Thompson loaded the pistol, Brookes endeavoured to wrest it from him and, whatever happened next, the 'contents entered the head of the poor man'. Somehow Thompson had been shot in the head. In a bloodied state, he repeatedly said, 'it was taken from me, it was taken from me', referring to the pistol and incriminating Brookes. With a ball shot lodged in his skull, Thompson had little to no chance of survival, and was taken to St Thomas's Hospital, accompanied by his new love, Sarah Tookey. Thompson died and Brookes was apprehended and brought before the Lord Mayor at Mansion House. He was remanded for trial the following week, while evidence was brought forward. *The Annual Register* reported that at the trial 'Brookes narrated the circumstances with a conscious innocence, and without the smallest appearance of fear or alarm for the consequences.' He was found guilty of manslaughter, but it might easily have been otherwise had it not been for his confident demeanour, social standing and friends. While a person convicted of manslaughter could walk free, a murderer would hang. Though killings, even fights between love rivals, might be considered manslaughter when there was no existing enmity and premeditation, the legal system allowed for discretionary leeway. John Brookes, honour still somewhat intact, resumed his business albeit with a decidedly more rakish reputation.[8]

In his dealings with his servants John was far from a model employer. An indignant Brookes had appeared earlier in court accusing his employees, a Mazarine Bell and his wife, of petty theft. They had, so he asserted, deprived him of a birdcage and a bag of feathers from his menagerie. His former servants decided to set the record straight with a rather embarrassing narrative. Brookes had neglected to pay them for a long time, and the Bells had been compelled to pay themselves from the money charged to those who wished to view the animals. From this they had secured a weekly wage of 14 or 15 shillings. Eventually they tired of Brookes's tardiness and moved on to different employment in May 1807. Months later, in November, Brookes suddenly noticed he was missing a birdcage and a bag of feathers, and he took his former employees to court for alleged theft. It transpired that was he in arrears for their wages and had lied about their requests for payment, but had met his former employees several times in the intervening months without making any allegation of theft or attempting to charge them. The court threw Brookes's flimsy claim out and the Bells were acquitted.[9] The magistrate and defence took this opportunity to remind John of another embarrassing transgression two years earlier, in 1805, when Brookes had himself appeared on charges of theft, charges that were substantiated. A stolen pug dog had been found in the possession of John Brookes and identified as belonging to a certain Colonel McKenzie. After two failed appeals and the fraudulent testimonies made by Brookes's servants at his behest, Brookes was held accountable for damages amounting to the not inconsiderable sum of £30. Clearly, John Brookes was a man often righteous in indignation and eager to get his own way.

In at least one instance John Brookes did get his own way, when he unscrupulously ruined a bird merchant. In 1800, a dealer in birds appeared in court desperate to plead for reparations for his ruined business. He had fallen behind on his rent and the landlord had seized his stock in lieu of payment. The dealer claimed that his stock of 200 pigeons, pheasants, owls and other birds was worth £64 10s. but it had been sold by his former landlord for a trifling £10 12s. in order to pay the rent. The purchaser and appraiser of the undervalued stock was Mr John Brookes, who had accompanied the landlord late one evening between 9 p.m. and 11 p.m. to see the birds, appraise them and agree on a price. Brookes was indeed killing two birds with one stone, taking on the not-disinterested dual roles of both appraiser and buyer. A fee for valuing the birds on top of steeply discounted birds: it is not only the early bird that catches the worm; Brookes struck late at night. The judge probed the dealer and Brookes about the actual value of the birds; the dealer protested that although the value of some of the birds had depreciated of late, his stock had been worth £64. Brookes argued that the stock really was depreciated and that he had valued and purchased them accordingly. After all, so he told the court, the demand for some varieties of birds was so low and the price of grain so unusually high that he had little choice but to set free his own stock of imported pigeons. With only his desperate daughter testifying to the value of her father's stock, the dealer was ruined when the court ruled against him. The landlord received his rent and Brookes, even as he positioned himself as a credible authority in court, aggressively purloined the stock of a competitor. As a family of animal merchants, the Brookeses were something of a behemoth.[10]

Gilbert Pidcock began his business on a small scale, touring a menagerie in the provinces. He had been active since the 1770s but by the late 1780s put down solid roots in London. Initially based at the Lyceum on the Strand, he later occupied premises on the Strand, at the Exeter 'Change around 1790. The Exeter 'Change was a large building with a shopping arcade where drapers, milliners and hosiers sold their wares; Pidcock took the apartments above these stores, advertising his animals on brightly coloured hoardings outside. He widely advertised the 'foreign beasts and birds' to be 'seen or sold' at his menagerie: a parade of elephants, lions, tigers, monkeys, bears, rhinoceros and vultures. The insurance policies that Pidcock took out from Sun Fire Insurance over a period of a decade record his material life. The properties and possessions of policyholders were typically insured for between £300 and £400 if they were tradespeople and from £500 upwards if they were 'gentlemen'. Goods such as paintings and china appeared insured separately from other goods, usually for sums between £30 and £50. Gilbert Pidcock was rather different.[11]

Pidcock's business premises were first insured for £250 and he insured his paintings, printed books, silver plate, glass, clothes and musical instruments for an additional £300. This included three organs, which must have been used when his travelling menageries were on the road. Pidcock's wild beasts and birds needed suitable accommodations and so the menagerie's dens, fittings and harnesses were insured for an additional £140. Already, we can see that Pidcock's business held insurance much higher than most policyholders with Sun Fire Insurance, and this is without consideration of his 'living birds and beasts & preserved specimens'. These were insured for £1,565, so large a sum that the animals were itemized in a catalogue lodged with the office of the insurer. Pidcock had good reason to insure his animal property against fire; when he kept his animals at the Lyceum on the Strand in the late 1780s, a fire had indeed broken out. At that time he was the proud owner of a 'beautiful male zebra' who was, apparently, so docile that the keeper could 'put young children upon its back'. In fact, the zebra, so it was claimed, had even been ridden from the Lyceum to Pimlico. Two stories circulated about the tragic death of Pidcock's male zebra at the Lyceum. One was that a mischievous monkey had got hold of a lamp and set fire to the straw on which he lay. The second story placed the blame on the keeper who, while going to warm a little milk for a kangaroo, left the zebra alone in an apartment.[12] He must have been tending to his kangaroo charge for a little too long since, while he was away, a tin lamp burnt through a hoop attaching it to a socket. The lamp fell to the floor and settled on dry hay, which burst into flames. Pidcock had purchased the zebra for £300, so it was a very expensive and distressing accident indeed. Likewise, other animals came with hefty price tags: a 'handsome male lion' cost £400, and a rhinoceros could be had for £700. When Pidcock's rhinoceros died of an inflamed dislocated leg in 1793 he tried to make good his loss by having the rhinoceros stuffed and placed on display. In 1800 a different rhinoceros of Pidcock's died, but fortunately, after he had sold it. It died in a stable yard on Drury Lane a mere two months after it had been sold to an agent of Francis II, the Holy Roman Emperor, for £1,000. It might take years to sell one of these animals and when people could not be persuaded to buy, at least the money garnered for entrance fees mitigated losses. People would tend not to part with their money to see a menagerie deprived of its choicest beauties and stalwarts.

Pidcock's menagerie grew in value, and by 1809 his insured value had increased from £2,400 to a hefty £4,000.[13] Contrasting his property with that of another animal dealer, George Kendrick, is instructive. Kendrick's business was at 42 Piccadilly, a stone's throw from the Strand. He insured his household goods, printed books, apparel and silver plate for £300. In addition his stock was valued at £400. Kendrick was doing comfortably well, but clearly in Regency London he was not at the apogee of the animal trade like Brookes or Pidcock.[14] Pidcock's colossal rhinoceros had come with an equally hefty price tag – the total value of Kendrick's prospering business.

The story of one of Pidcock's servants is one of a different kind of property. John Bobey, born to slaves in Jamaica in 1774, came to London aged 12 as a menagerie exhibit. He had been born a 'piebald' or 'spotted' curiosity, and his vitiligo made him an exoticized racial 'other'. One account said that he resembled 'a beautiful leopard'; parts 'black as jet', others 'equal to any European'. Bobey was purchased as an indentured servant for 100 guineas in 1789, at the age of 15, by Thomas Clark. Clark was a wealthy cutler and had grown enormously rich on his silverware business. He also sometimes bought and exhibited exotic animals, and so the young Bobey was put to work in the menagerie, where he tended to the animals and exhibited himself to the public. Later, Clark sold his animals to Gilbert Pidcock, who used them to bolster his menagerie. When it was his turn to be sold, Bobey protested. His servitude was sold for 50 guineas to Pidcock, and, as he was sold, Bobey is credited with protesting, 'I can't stand that, I will not be sold like the monkeys'. Black Georgians occupied a perilous position in Britain. Many lived in poverty if they were freed or freed themselves from servitude and slavery.

In 1771 James Somerset, an enslaved African due to be sold by a customs officer called Charles Stewart, escaped but was later recaptured and imprisoned on a ship leaving London for Jamaica. Somerset's godparents applied for writ of *habeas corpus*, a petition and summons for unlawful detention. In 1772 the landmark *Somerset v. Stewart* case heard at the King's Bench established that slavery was unsupportable in common law and, moreover, so too was it illegal to remove a person regardless of being a slave from England against their will. But the status of slavery as a whole remained ambiguous and persisted elsewhere both inside and abetted by the empire until the Slavery Abolition Acts of 1807 and 1833. Even the 'free' might be illegally recaptured and sold. Likewise, the 'freeborn' were subject to prejudice and kidnapping. Around 10,000–15,000 black people lived in London, mostly men, the majority living in the underworld of the city. In time, Pidcock released Bobey from his indenture and paid him a wage. Bobey soon left the menagerie to pursue an independent life, meeting and marrying an Englishwoman. With his new wife, Bobey established a travelling menagerie that utilized both his talents as an animal keeper and his appeal as an exhibit-turned-exhibitor. The Bobeys' travelling menagerie was a fairly substantial operation – a collection of 'monkeys, birds, and beasts' necessitating the 'keep of five horses and men'. A contemporary biographical account of Bobey's life supposed that 'by proper application of their savings' the couple would, with little doubt, accumulate a 'decent fortune'. Pidcock's indentured servant boy thus became a menagerist, achieving a degree of financial independence uncommon among black Georgians.[15]

Gilbert Pidcock died at the age of 67 in February 1810. His Grand Menagerie became the property of the Polito family, who renamed it the Royal Menagerie. Stephen, or Stephano, Polito was an Italian who became naturalized in Britain. He and his family had been the owners of a travelling menagerie for decades, and the Exeter 'Change was an opportunity to occupy a prime London location. Polito, as an impresario, beautified the menagerie with painted scenery and donated stuffed animals to William Bullock's museum at 22 Piccadilly. Showmanship and dealing in animals was a dual vocation. In fact, as a name, Polito's Menagerie lived on long after Stephen's death. Well into the 1850s menageries were travelling Britain and the Continent under the banner of Polito's Menagerie.

On his death in 1814 Polito left his menagerie to his brother, John, and money to provide for several daughters, Sophia, Phoebe and Antoinette. He left £1,000 to be given to them at the age of 21. In addition, he made financial provisions for his mother, still living in Molatirso on Lake Como.[16] His brother John married the daughter of Edward Cross, a man whose family name, as we have seen, had an association with exotic animals. Baptized in 1774 and living until 1854, Edward Cross was the last 'great menagerist' of the Georgian period. He had been superintendent of the menagerie at the Exeter 'Change for some 20 years by 1814, meaning that he had worked under Pidcock and Polito. From 1814 his rather pompously named Royal Grand National Menagerie exhibited and sold exotic animals at the Exeter 'Change on the Strand until its dissolution in 1829, upon which he established the Royal Surrey Zoological Gardens. Like Brookes, Pidcock and Kendrick, Cross was a policyholder of Sun Fire Insurance. His business was insured for £2,700.[17] Edward Cross had his portrait painted by the celebrated painter of animals, Jacques Laurent Agasse, in 1838. In rude health in his 60s, Cross is pictured in a stovepipe hat, cradling a lion cub in his arms. The Crosses remained in the animal trade into the late nineteenth century, with William Cross owning a dealership in Liverpool from the 1870s. From Liverpool William Cross orchestrated the import and export of wild animals and birds, selling them across the world as part of the well-established animal trade, with roots extending back to the mid-eighteenth century. From the world of the Georgian animal and bird dealers, we now turn away from their property to that of their customers. The animal merchants had been eager to secure and rear their property, and so too were their proud new owners. We move from the premises of the dealer, to the home of the owner.

CHAPTER 3

The Property of …

IN 1804 DR ROBERT John Thornton, lecturer in medical botany at Guy's Hospital and physician to the Marylebone Dispensary, bought a parrot from Brookes's Menagerie in the Haymarket. The blue and yellow macaw was rather expensive: Thornton paid 15 guineas, around £16, almost the annual wage of a domestic servant without lodgings. At the time he did not need to worry too much about money since he had a modest inheritance. Moreover, he hoped to make money on his botanical work *The Temple of Flora* (1799), his sumptuous homage to botany and Linnaeus, the master of classification. Thornton bought his macaw to grace his botanical exhibition at 49 Bond Street, displaying the showy exotic bird alongside his botanical specimens and floral portraits. The parrot was not particularly conducive to setting a genteel ambiance for the contemplation of Linnaean botany, nor the appreciation of floral portraiture; 'sulky' and 'unhappy' in a confined room, the parrot made 'those screaming noises so offensive in that tribe'. *The Temple of Flora*, for all its gorgeous plates, was a commercial flop; few copies of the expensive folio were sold and Thornton's exhibition soon closed. *The Temple of Flora* ruined Thornton financially and he never recovered the costs, but at least he had his parrot. The parrot came home with Thornton to Hind Street, Manchester Square where, according to *An Account of the Blue Macaw, the Property of Dr. Thornton*, the parrot flourished in a family home far removed from its 'years of servitude' in Brookes's Menagerie.[1]

At Brookes's the parrot had been tethered to a perch on a short chain and fed on a meagre diet consisting of scalded bread. Thornton took pity on his sulky parrot and 'from motives of humanity and out of experiment' he cut the chains that had confined the bird. The parrot's claws and feet were wasted and cramped; indeed the bird could not even walk, let alone fly, but rather 'tottered' around, clambering up and down the perch with the aid of his bill. In time, however, the macaw's feet recovered and 'his plumage grew more resplendent'. The raucous screams of discontent ceased and the parrot seemed, finally, to be happy. The family life of Thornton's parrot is an instructive peek into the private lives of these exotic animals after they left the premises of a bird seller or animal merchant and became the property of their new owners. The Thorntons took breakfast with their new parrot; the scalded bread was replaced with buttered toast, dumplings, potatoes, fruit and occasionally sugar. The parrot was trained to perch on the doctor's fingers and flap his wings, crack nuts and peach stones, and dance to a tune for a gathered company. Thornton, like many other Georgian parrot owners, was able to teach his parrot to talk. This particular bird was far from verbose yet mastered 'Pretty fellow', 'What's o'clock', 'Saucy fellow!', and 'Macaw', as well as the names of Thornton's son and daughter. Another charming, and perhaps sometimes annoying, habit of the parrot was his more than passable imitation of the knife-grinder who called on Manchester Square every Friday. The parrot's claws repeatedly scratching and rasping against his tin-covered perch imitated the sound of the grinding wheel turning against metal. The Thornton's parrot was settled into the daily routines of household life; the light of candles would awaken him and the state of his plumage was a portent of bad weather. The attentive Thorntons noted that their bird was 'an admirable weather gauge': 'in rainy weather the blue feathers look green, and also in clear weather when there are vapours in the sky'. The blue and yellow macaw was, to the Thorntons, something more than an expensive piece of property; he became part of their household.

Many professional city men from the middling ranks like Thornton were able to purchase exotic birds and animals; this interest was not just reserved for the aristocracy or gentry. Between

1 *Seaman Holding a Monkey*, 1816, Joshua Cristall
(1768–1847), watercolour and graphite

the 1720s through to the 1750s, when the naturalists Eleazar Albin and George Edwards wanted to draw curious foreign birds and animals from life, they visited the homes and premises of a whole host of Londoners. The footnotes of their works and the descriptions of their subjects are furnished with anecdotes as well as acknowledgement of the owners' kind permission to draw their property. The animals are never referred to as pets, instead, they are always described as 'the property of' their owner, just like Thornton's parrot. Many of these animals came from dealers in London, some were brought back from travel or residence in foreign climes and others still were sent as gifts from overseas. The wealthiest individuals could, like the Duke of Norfolk in 1789, commission an animal merchant to acquire a particularly desired animal. Brookes imported on his behalf 16 reindeer from Lapland; they arrived in Hull along with a consignment of Arctic mosses to help them acclimatize. The menagerists, animal merchants and bird sellers amassed their stock not only through their own agents or overseas travel but also though purchasing those choice birds or animals that came into ports in the hands of sailors. Seamen might sell on their property for a handsome profit, keep it for their own pleasure or give it to a relative. Londoners of more humble means, unable to spend cash on exotic luxuries, could thus sometimes receive them for free. In this way interesting birds and animals made their way into the homes and gardens of a variety of owners.

Albin and Edwards saw birds and animals across London and the city's environs. Sir Robert Walpole and his wife, for example, had kept a flamingo for some time, keeping it warm by the kitchen fire. Their household also included a scarlet lory and yellow-faced parakeet. The physician and collector Sir Hans Sloane had at his home in Chelsea a porcupine and opossum, amongst other creatures, including a crane that lived in his garden until it 'died by swallowing a brass sleeve button'. Sloane was followed around like a dog by his one-eyed wolverine and had at some point owned a marmot which he had fed poorly; the animal's teeth were so overgrown that they needed to be extracted.

Sidney Kennon, also known as 'Mrs Cannon', was midwife to Queen Caroline and delivered the future King George III. Her post put her in the position of delivering the babies of Georgian London's rich and famous, a job that in turn made Kennon a wealthy woman. Her house on Jermyn Street was crammed with natural curiosities, including shells and polyps, as well as anatomical specimens. In addition to parrots, Kennon had acquired a mongoose and marmoset. Neither was particularly well-behaved, the marmoset 'greedily devoured' a Chinese goldfish that it fished out of a porcelain bowl and the mongoose occupied itself by attempting to eat Kennon's birds. Other owners, like Thornton, were medical men or in allied medical trades. Mr Crittington, clerk to the Society of Surgeons, had a black lemur that he spoilt rotten with cakes, bread and butter, fruit, and vegetables. Thomas Walker, an apothecary in Crooked Lane, had a toucan with a taste for grapes. Walker would toss grapes to the toucan, and it would catch them 'most dexterously' before they hit the ground. A fellow apothecary, Mr Bradbury in Holborn, had a mongoose which he let loose in the house to clear it of mice and rats. Mr Scarlett, an optician near St Ann's Church in Westminster, kept a gerbil alive for several years. The gerbil would stay in its 'hutch' during the day but hopped around the room in which it was kept all night; the soft-furred rodent was difficult to handle and took to biting the hands of those who attempted to hold it. Proprietors of taverns and coffee shops owned exotic birds, too. Mr Bland, proprietor of the George Tavern on Tower Hill, for example, boasted two cassowaries and a vulture among his collection. Likewise Mr Mere, owner of a Bloomsbury coffee house, displayed his mynah bird on the premises.[2]

Some of the exotic animals that came to be the property of Georgian Britons were unusual or unconventional, even for the standards of the time. In 1793 the English naturalist Robert Townson travelled between Göttingen and Vienna on horseback, a journey that took him six weeks. But he was not travelling alone, for in his pocket in a little box was his 'favourite Musidora', a female tree frog or *Rana arborea*. Common in central Europe but absent in Britain, the 'beautiful little tree frog' charmed visitors to the Continent, who were particularly drawn to the frog's croak. Townson kept his frog in a glass of water in the window of his study and she became tame enough to be carried around in the palm of his hand. Sometimes she would fall from the windowsill and Townson would dust her down, returning her to the water glass. In warm rooms heated by stoves, Musidora lived through three harsh winters without the hibernation customary in her species. Townson's papers on amphibians, published in England, were replete with references to his tree frog. With some pride he wrote, 'I was able to feed my constant companion and favourite pet Musidora by hand with dead and living flies.' Townson was interested in the respiratory functions of amphibians and was experimenting in earnest on frogs, newts and even turtles imported from overseas for the purpose. He collected around 80 frogs in his cellar and kept them in a box. Some of the experiments were grim but Musidora was spared lethal and invasive experimentation. However, as part of his research, Townson did measure her water absorption rate and then drink the water she 'expelled', describing it as 'pure as distilled water'. In his published work Townson concluded his study of *Rana arborea* on a melancholy note: 'I am exceedingly sorry to terminate these remarks by informing Musidora's friends (for she had many) that she is no more. She sickened soon after she reached Great Britain, and died in the night of 25th of June 1794.' He attributed her death to exposure to the sea air during the crossing from Hamburg.[3]

The loss of an animal could sometimes provoke grief or sadness in their owners, and we will see in a later chapter how a certain amount of social criticism centred on the teary and inappropriate emotions felt, especially by women, on the deaths of parrots and lapdogs. As quoted in the prologue, Betty Snooke, the wife of the haberdasher James Hall, penned a witty but perhaps callous elegy on the death of her sister's beloved parrot. It was not only women who grieved at, or were aggrieved by, the sudden death or loss of a bird.

Whilst in Paris with Horace Walpole in 1765, the clergyman antiquarian the Rev. William Cole bought a parrot and noted in his diary, 'I made a purchase of a very beautiful grey parrot, who sung and talked the best I ever heard.' Cole bought the parrot and his cage for four guineas, and it accompanied him on the long journey back across the Channel and on to Cole's rectory at Bletchley, Buckinghamshire; he bought two pence worth of sugar en route as a treat for his new companion. Unfortunately, less than a year later, the parrot died unexpectedly and Cole's gloomy diary entry for 25 January 1766 read: 'Foggy. My beautiful parrot died at ten last night, without knowing the cause of his illness, he being very well last night.' A few years later Cole's aristocratic patron gave him a gift, 'a beautiful red and green parrot'. Cole might have reasonably hoped to have better luck second time round, but this was not to be – his maid Molly carelessly let the bird fly away.[4]

The death of a foreign bird or beast often marked the end of one sort of property ownership and the beginning of another; their owners might send 'dear Poll' to a taxidermist or pass her on to an anatomist for dissection. The aristocracy and gentry often bestowed their property in death upon a museum collection as a fine way to boost status as a patron. Certainly some shop owners, wanting to make good on the loss of their parrots, had them stuffed and placed in the window of their shop. Georgian London, by the 1780s, had no shortage of taxidermists turning a fine trade on the fashionable trend for natural history. Their trade cards assured their customers that the birds and animals would be 'preserved in any attitude like life'. John Chubb, 'stuffer and preserver', bought and sold animals both living and dead. Thomas Hall was a particularly renowned taxidermist, operating from the 1780s at City Road, Moorfields. Hall reassured his clients of his ability and sensitivity, 'as there are many ladies and gentlemen who are partial to their birds and favourite animals', he promised to preserve them 'so near to life to be scarcely distinguished' from the living. Moreover, so he claimed, the stuffed and preserved remains were 'warranted to last beyond expectation'. Parrots, monkeys and other expensive exotic favourites could be given a second lease of life. This trend perhaps reflected not only fine sensibilities but also a more pragmatic attitude towards property. Costly and unusual possessions ought not to be so easily thrown away. Likewise some owners would not so readily allow their investment to drop off the perch to dust and grubs. If a parrot did not live as long as expected, it might at least 'last beyond expectation' as drawing room taxidermy.[5]

Loss might take a different form, as poor William Cole well knew. When eighteenth-century Londoners lost their horses or dogs – or, as was sometimes the case, had them stolen – they could post in the classified section of one of the city's many daily or weekly newspapers. In some of these lost-and-founds owners sought the return of a prized parrot. When Mrs Man, a proprietor of a coffee house, lost her parrot in May 1706 she hoped somebody had found it after it flew away and 'strayed over some of the neighbouring houses'. Some, like Captain J. Lesty in June 1707, also placed multiple advertisements in the *Daily Courant* and offered substantial rewards. Lesty's small green parrot had flown out of his cage and out of the window of the house on Church Lane, Limehouse. Lesty knew that a parrot flying around East London would not have gone unnoticed. The person that had 'taken it up' was encouraged to bring it to the captain and receive a guinea for their troubles. Perhaps the captain's parrot had flown off whilst its seed was being changed or water bowl refilled; another parrot that appeared in the *Daily Courant* in 1716 was stolen out of an entry to a house in Whitehall. Whoever had 'taken up' the said parrot was promised that they would be 'very well-rewarded for their pains'. They could return the parrot to a stationer near, appropriately, Scotland Yard Gate who was presumably dealing with this matter discreetly on behalf of the parrot's well-heeled owners. This sort of theft was not uncommon in Georgian London; thieves often stole items of high value from named individuals with a view to returning them later for a reward.[6]

Sometimes the theft of exotic birds led to arrests and court cases, especially when the bird could be correctly identified and witnesses produced. In a strange turn of events, the men who had sold the birds could appear in court to testify. In December 1768 a man named William Enoch stood in court charged with the theft of two gold

China pheasants, which he claimed he had caught in the wilderness by a roadside near Highgate. The birds had been stolen from the menagerie of the Hon. Charles York: the door had been forced open with an iron instrument and the birds snatched. The birds were presented in court as evidence and both York's servant and gardener vouched for the birds: 'These are the same, I know them well,' 'I am very sure these are the same produced.' The defendant, Enoch, had attempted to sell the pair of pheasants at Newgate Market. The prospective buyer became suspicious and, doubting the provenance of the birds, he took them to a dealer in birds in Holborn, Mr James. The dealer had been told of the theft and recognized the birds as those belonging to York. The birds were valued at three shillings a piece, meaning that the pair was equivalent to a labourer's weekly wage – a tempting target indeed for a thief. In fact, this valuation was almost certainly a rather substantial underestimation of the market value of the exotic birds; the value of stolen goods was increasingly underreported in Georgian Britain. At that time capital crimes included some that by definition are characterized today as somewhat petty crimes. Larceny, theft of property, was a capital offence and in playing down the value of goods the court could sometimes subvert the otherwise mandatory death penalty with a discretionary alternative. Enoch was found guilty and transported to the American colonies.

Many judges and jurors were reluctant to condemn defendants to death, and transportation was seen as an alternative, a useful middle ground between the harshness of the death penalty and the perceived leniency of imprisonment, fines or whipping. The public perception of transportation to a distant land was one of a fate better than that of the hangman's noose but infinitely worse than imprisonment. The Georgian state sought to enshrine in law a robust defence of property and thus, naturally, the interests of the propertied. The have-nots, even in the most impoverished of states, were supposedly deterred from stealing property by punitive sentences.

Later, in 1789, Thomas Andrews was indicted for the theft of another pair of gold China pheasants, this time belonging to the

Member of Parliament Jervoise Clark Jervoise Esq. The birds had been stolen from his aviary and somehow during the course of the theft the cock pheasant's gorgeously opulent tail feathers had been pulled out and inadvertently left behind as criminal evidence. Andrews took the pheasants to Gough's Menagerie on Holborn Hill and tried to sell them to the proprietor, William Gough. Andrews had, somewhat suspiciously, turned up at the menagerie with a pheasant stuffed unceremoniously in each of his coat pockets. With two writhing and squawking birds in his coat, Andrews probably cut an odd figure when he walked into Gough's Menagerie. Pulling them out for inspection, Gough detained Andrews, believing them to be stolen property. Andrews had offered him the pair for 30 shillings; the court valued them at 20 shillings. The missing tail feathers were Andrews's downfall; he must have rued leaving them behind. Jervoise's gardener, George Rolls, appeared in court to vouch for the birds, with the feathers. He also recognized the burn marks he had made under their left wings. In the decade or so between the thefts of the two pairs of Chinese pheasants, the situation had changed, so that those sentenced to transportation had actually spent their sentences confined to hard labour on rotting prison hulks moored on the Thames estuary. War with the now-independent American colonies from 1776 closed off transportation as a judicial recourse; it was not until 1787 that transportation resumed again, this time to Australia. Thomas Andrews, sentenced to transportation in 1789, would have been on the second fleet to New South Wales: half a dozen ships with settlers, convicts and supplies sent to Port Jackson.[7]

The theft of Lady Harriet Gill's parrot attracted a great deal of attention, not on the merit of a parrot theft being anything unusual but rather because the crime and its aftermath brought two aristocratic households, along with the parrot, to appear in front of a magistrate on Great Marlborough Street. This bird, bestowed with the tongue-in-cheek sobriquet 'the parrot of contention' in periodicals during 1802 and 1803, was otherwise known simply as Poll. Poll had been given to Lady Gill some years earlier by Earl Wigton and was stolen from the entrance of her house on Wigmore

Street, Cavendish Square. The thief was evidently gutsy and adept to make away with the bird, cage and all, from a West End aristocratic residence full of servants. The loss of the parrot was advertised in newspapers as well as bills posted, but there were 'no tidings of Poll' for a fortnight until a servant of Wigton happened to pass through Henrietta Street, near Covent Garden. There, in the house of the Countess of Granard, was a parrot swinging in a cage that was, so she asserted, the very parrot given to Lady Gill.

The Countess of Granard's cook took Poll to the magistrate's office, where Wigton and his servants swore that this indeed was the stolen bird. Wigton tried to prove the ownership of the bird by putting his hand in the cage and tickling the bird; the bird bit him and made a croaking sound just as Wigton said it would. Granard's cook was not impressed: 'Excuse me, my Lord, any parrot will do that when you hurt it so,' she said. Not to be beaten, Wigton then tickled the bird's breast and elicited a 'Boo! Boo!' just as he had promised the magistrate. The household servants of Granard claimed that Poll had been given to their mistress by Lord Berkley in 1793, and that she had kept it in their house a full ten years before she left for France, entrusting her cook with its care during her absence. The Countess, it was claimed, would not part with her bird for even £50, to which the opposition replied that Gill would not part with Poll for double that sum. Gill's parrot, normally so talkative and prone to vulgar swearing, was unusually taciturn in court. That was except for the comic relief Poll had provided when the magistrate called for her to be brought into the court by the cook; the room resounded with 'No, No,' and the laughter of those assembled. Faced with a welter of evidence and the insistence of witnesses on both sides, the magistrate had little recourse but to wash his hands of the two feuding households. Instead, he reserved judgement and expected Earl Wigton to call upon the Earl of Granard and clear up the matter like gentlemen. The strong feeling of ownership over Poll went beyond a defence of property, and instead was bound up with the emotional attachment the two women had to their bird. Perhaps Gill's Poll really had been stolen and found her way into the Granard household. Or perhaps a more reasonable explanation is that it was a case of mistaken identity and Granard's cook would not budge, knowing that on her return from France her employer would vent her wrath.[8]

In October 1802 the 'celebrated Miss Poll' died at her home in Half-Moon Street, Piccadilly. Miss Poll was upwards of 30 years old when she died. She had been born in Bristol and was praised for her musical memory and fine singing voice. Her morning levees had been attended by 'people of the first rank and fashion' who came to listen to her sing. If the assembled company were impertinent enough to interrupt her performance, she would stop as if to demand the 'utmost silence'. Miss Poll, a green, grey and red parrot, was first the property of Dennis O'Kelly, who made his fortune on racehorses and became a substantial landowner with a Middlesex estate and London townhouse. O'Kelly and Charlotte 'Mrs O'Kelly' Hayes did not have auspicious beginnings: Dennis had been a gambler and served time in a debtors' prison; Charlotte was a prostitute-turned-brothel-keeper. Yet, on the back of the wealth generated by studding winning racehorses – their first horse being purchased on the proceeds from Charlotte's brothel – they became incredibly wealthy. At some point around 1770 Dennis spent 50 guineas on Miss Poll, though some said he paid 100 guineas. The bird was believed to be the first parrot to be born in England and Dennis paid a woman to convey the bird safely to him from Bristol to London. The prodigious bird was described as one of Charlotte's 'greatest distractions'. Miss Poll was well known in her time as possessing a range of 'astonishing qualities' by any measure, being a talkative and responsive parrot. She would answer questions and memorize melodies, engaging in banter, albeit limited, with the audience. The fine dining and drinking caught up with Dennis O'Kelly and he died of gout in 1787. His obituary in the *Gentleman's Magazine* made more mention of his parrot than of his own life. Miss Poll was left in his will to Charlotte, though actually the parrot ended up being passed on to his nephew Colonel Andrew O'Kelly. This gentleman was known for his urbane manners and was well liked in society, a state no doubt helped along by his willingness to host those who called upon him with a pleasing 'parrot levee'. The bird's performance was certainly impressive. One guest, a Rev. W.H. Herbert, recalled seeing 'him' in 1799 turning around on the perch, singing songs and beating the time with 'his' foot: 'it articulated every word as distinctly as a man could do.' Moreover the parrot would finish singing any song, from a repertoire of around 50, that a person, would hum to it.

Miss Poll might really have been a 'he'; 'he' was taught to sing and speak by the woman who had sold 'him' to Dennis O'Kelly in the 1770s, and it was alleged that the parrot's voice so resembled 'the female voice divine' that passersby in the street thought the parrot was a woman. At least several sources thought Miss Poll to be a male parrot with a woman's voice. On the days of the levees or 'parrot concerts' Half-Moon Street was 'filled with carriages and an admiring crowd', rendering the street impassable. Colonel O'Kelly was offered a large amount of money for Miss Poll, with offers of 100 guineas coming from, allegedly, Gilbert Pidcock, who wanted to exhibit the bird in his menagerie. O'Kelly would not consider even 500 guineas. The colonel, 'out of tenderness' and concern that the bird might be 'put to work too hard', did not part with his bird until that sad morning in October 1802 when she, or perhaps he, died of a 'bloody flux' – some sort of dysentery or passing of blood. The death of the bird that O'Kelly had inherited from his uncle was said to 'have thrown the West-end into condolence and confusion', having been 'the wonder of London for years'. Miss Poll was dissected and stuffed by Dr Kennedy and the anatomist Joshua Brookes, son of the bird and animal merchant Joshua Brookes.

They were looking for some kind of anatomical basis for the bird's unusual talents; what they did find was very well developed vocal cords, attributed to the regular exercise of the 'parrot concerts'. At first the newspapers reported that no cause of death had been discovered, but a later second discovery settled the matter of the parrot's sex and the cause of the fatal 'bloody flux'. When Brookes auctioned and sold his anatomy collection in 1828 his catalogue listed for sale the body of the 'universally known' and 'celebrated parrot' of the late Colonel

O'Kelly. The preserved parrot's oviduct contained a 'defective egg'. Poll was indeed a 'miss' and had died because she was egg-bound.[9]

The inclusion of exotic animals in wills was motivated by a desire to protect property with financial and emotional value. The clergyman naturalist the Rev. Gilbert White, for example, inherited his tortoise Timothy from his aunt. The bequest of exotic animals was probably more commonly an ad hoc verbal or tacit agreement. That the inclusion of parrots in wills attracted the attention of the print press suggests this was an uncommon practice, and indeed one that raised eyebrows. In 1762, Mrs Killinghall, the widow of a surgeon to the naval docks at Chatham in Kent, left an annuity of £5 per annum to her parrot and cat. This was certainly more than enough to keep these two animals in comfort, yet even this sum paled in comparison with that left by Elizabeth Orby Hunter to her parrot in 1813. When Orby Hunter, widowed since 1791, died at the age of 60 she had lived with her parrot for some 25 years, having bought the bird in 1788. Orby Hunter and her parrot had been through a great deal together, not least when they lived in Berlin during the Napoleonic Wars. Mrs Louisa Adams, the wife of American diplomat and future president John Quincy Adams, had met Elizabeth in Berlin and described her as 'eccentric', smeared in rouge and a hysteric prone to tears. This notwithstanding, Orby Hunter did maintain some sort of social circle beyond her parrot; she died at eight in the evening at her home on Upper Seymour Street, Portman Square, and had been due to have a 'fashionable party' at ten. She went to considerable pains to provide for her parrot after her death and strenuously stipulated her wishes in her will. The bird was bequeathed 200 guineas a year to be paid during its lifetime to the person who cared for it. The parrot was to be left first in the charge of Mrs Mary Dyer, a widow living in Park Street, Westminster. To help settle the parrot in, Dyer was to be given the sum of 20 guineas to buy a large and tall cage to accommodate the bird – the parrot must have had handsome lodgings indeed. Orby Hunter had thought, or perhaps over-thought, through a host of contingencies: Dyer could pass on the annuity and the bird to another person but they could not be a man, a servant or a foreigner. She did not want her bird to leave the country or to be killed by a mercenary servant in need of cash. Her parrot was to be kept by an English lady of means (a 'respectable female') so long as it lived. Moreover, the very same parrot would have to be shown to the executors in order to receive the monies, thus pre-emptively foiling any sinister switch-the-parrot plots. If any of the individuals to whom Orby Hunter had left legacies sought to contest her will on grounds of the parrot, her executors were instructed to forfeit them their share. She justified her reasons in bitter terms: 'I owe nothing to anyone, many owe me gratitude and money, but none have paid me either.' The parrot presumably lived to a ripe old age on a generous annuity.[10]

Exotic animals were, then, valuable commodities and property in eighteenth-century London, worth stealing, worthy of stipulating in wills, going to court over, and even bequeathing money to. Parrots, perhaps the most common household exotic, could become something more than simple property, becoming familiar members of a household and even attaining – like Miss Poll – a certain celebrity status. Those ladies and gentlemen mourning the loss of their favourite could call upon the services of a taxidermist to preserve their property, hinting at a degree of sentimental value not easily quantifiable in monetary terms. Stuffed skins, of course, were still valuable and collectable in a period when natural history was both a fashionable pursuit and an intellectual undertaking; the desire to hold onto one's exotic animal or bird in death was probably motivated by financial considerations as much as tender feelings. As valued commodities, however, exotic animals or rather products made from them were more often consumed as ingredients or luxury goods in Georgian London. Under the theme of 'Ingredients', the next section reveals the use of turtle, bear grease and civet as a high-profile and controversial sort of consumption in the long eighteenth century.

PART II ᵜ: INGREDIENTS

ANIMAL MERCHANTS SOLD THEIR animals as living commodities, but the consumption of exotic animals by the elite was also gustatory and cosmetic. Bear grease, civet and turtle were highly desirable ingredients and consumer goods. In this way, exotic animals permeated the material lives of the higher echelons as consumed edibles, olfactory pleasures and hair products. Once processed as ingredient or product, this consumption was of course markedly different from that of the stock of the animal merchant. The former were valued as exotic goods because they could be materially transformed into a secondary product of greater value. This conspicuous consumption was the source of a great deal of cultural anxiety in the Georgian period. Luxury and excess might threaten the moral fabric of the people or spoil good English taste. Under the theme of Ingredients the following three chapters explore the production, circulation and cultural reception of these exotic ingredients.

CHAPTER 4

'Turtle Travels Far'

W HEN THE PHYSICIAN AND poet John Armstrong wanted
to express his admiration for the libertine radical political
figure John Wilkes, he used the taste of turtle. Although this
admiration would later turn to disdain, the turtle allusion would
have done double duty as a jocular quip-turned-barbed jibe. In his
epistolary poem *A Day*, dedicated to Wilkes, Armstrong describes
the delicious things that Wilkes, newly returned from his banishment
in France, would eat: veal or even a roast hare, the hare's 'generous
broth I should almost prefer to turtle soup, though turtle travels far'.
Georgian literary readers would have understood the connotations
of this literary flourish: the exotic turtle did indeed travel far.

The taste for turtle flesh was a prominent and controversial 'mania'
in the latter half of the eighteenth century. Armstrong chose the
delicacy that was at the apogee of Georgian fine dining, a dish second
to none. However, he was not simply alluding to the taste or even the
nautical miles that it took to bring turtle to Britain. As this poem
was read over the following years, he would also turn his readers'
minds to the political controversies of turtle eating, not least because
Wilkes was later an alderman and Lord Mayor of London (1774).
As we shall see, the aldermen were a group that particularly came to
be associated with gluttony. As such, Wilkes was intimately linked
to the turtle-guzzling reputation of the Corporation of London.
Moreover, even as Wilkes was imprisoned in King's Bench Prison,
he had eaten two turtles in his first month after being sentenced. The
turtles were gifts from wealthy allies.

In this chapter's biography of the turtle we encounter the turtle in ships, taverns, in print, on dinner plates and on the stage. Turtles certainly travelled far in Georgian culture because they acquired meanings and values well beyond their taste. We start first, however, with the journey, tracing the travels of the turtle from the high seas to British ports.

The taste for dressed turtle and turtle soup, sent to the table in a baked turtle shell or a soup tureen, attracted considerable attention as a culinary affectation of the privileged who eschewed good beef and mutton for the gustatory pleasures of turtle. Before turtle was a elite delicacy, it was, however, a staple food of sailors in the tropics and eaten almost daily by English settlers in the West Indies. The Georgian mania for turtle can, in part, be attributed to the gentry who owned or administered plantations and colonial operations in the Caribbean. Eating plentiful turtle on a regular basis, they developed a taste for it and took it home.

The turtle trade was not a mere trickle of turtles into the ports of Britain; astonishing numbers of turtles were involved. In tropical waters, large quantities of turtle could be brought aboard as cargo in a single day. Some of these turtles weighed as much as 400 lb each. Turtles could be caught with ease on beaches at night, when they came to lay eggs and lumbered clumsily across the sand. In shallow waters, turtles could be lanced with spears and hauled aboard. The largest required winching aboard with ropes attached to their shells. In 1709, anchored in the Galapagos, the crew of the English ship *Duchess* caught 150 turtles in a single day. It was not always necessary, however, to go to strenuous efforts to acquire a supply of turtles. Almost effortlessly they could be procured from local peoples through trade and barter. In 1770 whilst HMS *Endeavour* was anchored in waters off Indonesia, locals approached in boats 'full of all manner of live animals for consumption and trade; – turtle, fowls, ducks, parrots, rice-birds, monkeys, and other articles'. Lieutenant Cook was able to buy a small 6-lb turtle for one Spanish dollar. Only the larger, robust and profitable turtles made it to Britain, so we can suppose that Cook's small turtle the size of a dinner plate was swiftly dispatched to the care of the ship's cook.[1]

The shipping conditions during the passage to England were not optimal for securing the highest-quality turtle meat. As the hull of a ship lurched with the waves, the chelonian cargo suffered. Ominously, we read that the 'workings of the ship occasions them to be beat against the sides of the boat'. Cracked shells and battered flippers aside, the taste of the meat was also rendered less palatable.

Cook and his officers were able to 'fare sumptuously every day' on servings of turtle, remarkable not only for its plenitude but also for its flavour. They agreed that 'they were much better than any we had tasted in England,' and attributed this to the turtle being 'eaten fresh from the sea, before their natural fat had been wasted' and their 'juices changed by a diet so different from that the sea affords them'. This diet was 'garbage', presumably the scraps from the officers' table or paltry sailors' rations. It was a far cry from the sea grasses, seaweeds, sponges and jellyfish to which sea turtles are accustomed. Confined in a wooden tub and subsisting on garbage, it is a small wonder that the turtle's flesh wasted and became a less agreeable delicacy.

Since turtle travelled a long way to its detriment, the parson-naturalist the Rev. Samuel Ward, with a dash of tongue-in-cheek humour, opined that it would be better to reverse the state of affairs in the turtle trade: 'instead of bringing the turtle to the epicure, the epicure ought to be transported to the turtle.'[2] Surprisingly, London's epicureans did not eagerly take to the seas, and so it remained that the best turtle was to be found served on a captain's table and not in the city's fashionable dining rooms or gentlemen's clubs.

Certain species of turtles appealed to the palate more than others. Although hawksbill and loggerhead turtles were imported for their flesh, they were deemed a poor relation to the gastronomically appealing green turtle. Ward described the meat of most sea turtle species as 'rank' or 'indifferent' and reserved praise for only the green turtle. Apparently, this was a particularly easily digested, delicate and nutritious meat. As a material for crafting goods, though, the thin shell of the green turtle was more suited for inlay than more substantial working. Instead, it was the more robust shells of the less appetizing 'rank' turtles that were more highly prized by artisans in the creation of boxes, cabinets, combs, spectacles and other trinkets.

For many of the turtles that were shipped into England, their long nautical voyage came to an end on the quays and docks of London. In the pages of London's advertisers and gazettes turtles were regularly advertised as part of the exotic cargo pouring into the city's eminent grocers: pineapples, limes, tamarinds and turtles. However, from the docks many made their way to the specific warehouses and shops of the turtle trade. The coffee houses and grocers of Covent Garden specialized in the retail of live turtles; Ward's Original Turtle Warehouse regularly imported turtles from the Bahamas, boasting a 'great variety' to customers. Likewise the Queen's Arms Tavern in St Paul's Churchyard exhibited and dressed large numbers of turtles. Similarly, Wood's Coffee House insisted, 'families in town and country may have any quantity of turtle in the highest possible perfection.'[3] Other premises too advertised their willingness to dispatch turtles to the countryside, alive or dead. Some dealers of turtle were also eager to buy live turtles from private individuals returning from the tropics, no doubt profiting handsomely through resale. Turtles that were peculiarly noteworthy on account of their large size were advertised prior to their sale to ensure a high price from single or multiple purchasers, though many other smaller turtles awaited the inspections of a cook in anonymity. A pound of turtle was typically priced between one and two shillings, which was equivalent to the daily wage of a manual labourer.

Like many of the city's service industries catering to the elite, the turtle trade thrived on the London Season. Along with the seasonal sitting of Parliament between November and June came a coterie of politicians, courtiers, officials and their households into the city. The turtle merchants 'continued dressing all season', which is to say they had established reliable turtle supplies to replenish a stock for the dining table. These merchants certainly had enough incentive to secure a steady supply of turtle, it being a desirable ingredient and dish of choice at the dining table when the aristocracy and elite resided in the city. Such conspicuous consumption of large quantities of turtle in gentlemen's dining clubs and in fine dining rooms engendered satire, humour and censorious cultural criticism. Indeed, the consumption of turtle brought to the foreground of contemporary minds questions of masculinity, political dignity and gluttony.

Once in the kitchen, turtles were kept alive until the night before they were on the menu. Larger turtles would be hung up on ropes with their flippers tied, lest they thrash and 'be troublesome to the executioner'. The process of dressing a turtle for eating was a grisly and protracted process. One such method for preparation was described in *The Ladies' Assistant for Regulating and Supplying the Table* (1787). First the turtle would be upturned and decapitated, scalded, and then hung to drain of blood. Turtle soup could be thickened with flour, sweetened with butter, and then flavoured with onions, thyme, parsley and a touch of hot cayenne pepper. A more substantial meal might be rustled up by first coating a turtle shell with flour, before filling it with turtle flesh brushed with egg. This turtle-stew-in-a-shell would then be baked until browned. Hannah Glasse's *The Art of Cookery Made Plain and Easy* (1747) contains a turtle stew recipe that – in addition to ½ lb of butter – called for 1½ pint of Madeira wine, with a cooking time of 'four or five hours doing'. More parsimonious cooks might supplement turtle meat, or even substitute it entirely, with 'mock turtle', a boiled calf's head separated from the bone. Mock turtle became regular aspirational fare at the tables of the middling-sort.[4]

Quite why turtle flesh was such a taste sensation is not entirely clear. Gourmands characterized the greenish-hued flesh as sweet and buttery, but others found it unpalatable. That turtle travelled far to the table and was such an expensive ingredient must have surely fuelled a taste for turtle amongst those who would have otherwise turned their noses up at the pungent flesh. However, given that aristocratic men in dining clubs also conspicuously consumed turtle, an element of epicurean machismo accounted for its popularity, too. Gorging on turtle became akin to generous portions of patriotic beef, something *English* and *masculine*. In a fictional account of a turtle feast, a 'haunch of venison' sat 'untouched', passed over in favour of the turtle imposter. The consumption of too much also went hand in hand with waste.

Eating a large turtle straight from the shell was a sociable affair. Some of the turtles consumed collectively at a feast were huge – up to 400 or 500 lb in weight. This was quite of feat of gluttony, indeed the character of the gluttonous city alderman emerged as a figure of satire in the late eighteenth century. A 'turtle feast' might entail polite enjoyment of a delicacy, or a Bacchanalian affair with 'saws, chisels, and instruments' employed to scrape every last tasty morsel from a carapace. One wit declared that it was not safe to attend a 'turtle feast at one of the city halls, without a barrel-hilted knife and fork'; that is, cutlery equipped with a cage to protect one's hand, like a sword. So voraciously did turtle-eaters brandish their knives and forks that diners needed to protect not only their hands but also their plates of turtle.

In contrast to the 'turtle feasts' of the city's aldermen, the dining club of the Royal Society hosted 'turtle dinners'. The phrase hints at the strenuous efforts taken to distinguish the polite from the vulgar. After attending a turtle dinner at the Royal Society on 27 October 1785, James Watt, pioneer of the steam engine, wrote to his wife in a reassuring tone that the gentlemen of knowledge had remained polite and dignified: 'never was turtle eaten with more sobriety, and temperance, or good fellowship'.[5] This was perhaps not always the case. A turtle weighing 400 lb was dressed at the club in 1752. The dressing of this turtle – the slaughter, removal of the internal organs and preparation of the flesh – was a tremendous undertaking. The dining club's kitchen proving insufficient for the task, 'the company were forced to dine in a room different from what they used to dine in.' The turtle's evisceration had commandeered the use of the dining room.

So while gentlemen ate turtle with sobriety in amiable fellowship, London's aldermen had become synonymous with gluttony in the public image. The Corporation of London's aldermen were elected for life as officials, based on property qualifications, to form a close-knit oligarchy in the city's governance. With constant accusations of misconduct, corruption and unseemly living, London's publications depicted the alderman as a portly diner with gusto. The alderman was particularly associated with his favourite dish, the turtle. A story written at the beginning of the turtle craze in 1755 appeared in the *London Magazine*, which sought to give 'readers an account of a new species of luxury eating'. Readers were drawn into a debauched world of turtle eating, where men robed in loose-fitting garments gorged to excess, monstrously devouring flesh. 'If you love turtle, I'll show you a sight'; and what a sight – the walls of the dining room were lined with shells, 'the trophies of his luxury'. And in the kitchen, six turtles swam in a large cistern awaiting their gruesome fate.[6]

Our guide to the hidden world of the turtle feast ponders the nature of the invitation to feast, an offer that he declines. Recalling the rampant speculation of the South Sea Bubble of the 1720s, the *London Magazine*'s writer opined, 'I consider these invitations to have more real value than so many shares.' The equation of turtle with over-valued shares went further: 'it will soon be as common to transfer shares in turtle, as in any other kind of stock.' Stock in the South Sea Company had rapidly inflated, and many had been caught up in the wave of greed. Eating turtle in Exchange Alley, the city's bankers and goldsmiths were foolishly succumbing to greed.

Stockjobbing bankers and goldsmiths, many of whom were councillors or aldermen, were a common target of ire in print culture. In the Georgian period London's high bourgeoisie – bankers and merchants – were increasingly purchasing land and marrying into the aristocracy. The city's aldermen were typically drawn from these sorts of mercantile families. This gentrification of the bourgeoisie was a source of anxiety and rich satire. William Hogarth's *Marriage à-la-Mode* famously depicts a marriage between the daughter of an affluent ambitious alderman and the son of a bankrupt aristocrat. In the exchange of title for money Hogarth critiques luxuries, taste and sexuality. Gorging on expensive turtle and swallowing up aristocratic families on the decline, the city's merchants and bankers were targets ripe for parody.

A satirical alderman character known as 'Mr Turtle' appeared in news columns and editorials. Readers of these print satires must have smirked knowingly at those stories that poked fun at the city's officials. The phrase 'turtle feast', alluding to excess and greed, appeared in the droll simile 'as gay and lively as an Alderman to a turtle feast'. Indeed, when addressing the Court of Aldermen during the lord mayoral election of 1773, a Captain Allen forcefully addressed his esteemed audience as a 'set who cared for nothing but gorging turtle, and smacking their lips'. The lip-smacking alderman clearly tickled the sense of humour of the caricaturist William Dent, whose 'City Coalition' (1783) showed a stout alderman dressed in his finery kissing the turtle lovingly embraced in his arms. Similarly, an earlier etching from 1776 pulled no punches with its stark waggish title, 'The English Glutton'.[7]

The turtle-guzzling alderman appeared in other areas of print culture too; at least two plays from the 1760s featured a turtle guzzler enjoying the pleasures of the fashionable spa town of Bath. In *The Widow'd Wife*, a comic play that first ran at Drury Lane Theatre in 1767, the 'luxuriously living' Alderman Lombard becomes *au courant* with the happy news that 'increased importation of turtle' to Bath provides a 'good excuse' to visit and ease one's gout in the warm waters of the spa. The dramatist Samuel Foote's play *The Patron* was first performed at the Haymarket Theatre in 1764 and is bursting with topical turtle humour designed to both lampoon and delight. Sir Peter Pepperpot imports a 'glorious cargo of turtle' to London and stands to make a fortune from selling his nine turtles. Seven of the nine turtles are sold to the London aldermen for corporate entertainment, with each alderman allotted a sitting of 6 lb of turtle, and their wives a 5-lb sitting. These ample portions are a contrast to the fate of the remaining two turtles, 'sickly' after their sea voyage. These turtles were dispatched to far-flung Yorkshire to less discerning buyers: 'Ay! What, have the provincials a relish for turtle?' Sir Pepperpot has been able to profit handsomely from the gentry of Yorkshire's newly acquired taste for turtles. Pepperpot gleefully quips 'Sir, it is amazing how this country improves in turtle and turnpikes.'[8]

Foote's satirical comedy was keenly observed, since during the 1750s and 1760s a 'turnpike boom' revolutionized road transport in the Georgian period. In these two decades some 300 turnpike trusts were established and maintained or constructed 10,000 miles of road, paid for by tolls and subscribers. This led to shorter travel times and a lower cost of transporting goods. This notwithstanding, London to York was still a 70-hour journey in the summer – 90 hours in the inclement English winter. The turnpike road network helped Pepperpot's sickly turtles reach Yorkshire. Using improved roads, the turtle boats that arrived in Bristol were also able to supply their delicacy to consumers relatively far from that port.

Bristol's status as a major port and Georgian England's second city supported a thriving turtle trade. As early as the 1770s numerous taverns and turtle warehouses in Bristol were open for business – premises such as the Old Turtle Warehouse on Corn Street, or the Bush Tavern. Turtle was popular in Bristol; in the summer of 1776 John Weeks, victualler at the Bush Tavern, was serving five shillings a head 'turtle ordinaries' – set-price meals – on Tuesdays, Thursdays and Saturdays. Throughout the 1780s and 1790s, the Bush Tavern regularly advertised in the *Bath Chronicle* to attract the custom of the elite consumers in that city, 12 miles away. In fact, in 1796 a total of three Bristol taverns announced they would be 'dressing' fresh turtle daily all season. In the 1780s, on a turnpike road, in fine weather, the journey from Bristol to Bath was a mere one to two hours by coach. Live turtles could be purchased in Bristol and dispatched to the country in haste. Somerset's well-to-do could order turtle in the morning and serve it for dinner, because it was close enough to Bristol.[9] Alternatively, turtle could be bought prepared and potted to last up to ten days in transit. Starting in Bristol, potted turtle dispatched to London by mail coach might arrive in as little as a day. Even in a bad winter, ten days was sufficient to carry turtle as far as Edinburgh.

Bristol's aldermen also had a particular liking for turtle, and modifications to the Civic Mansion House in 1786 necessitated the payment of £4 for the installation of a 'large turtle tub'. In 1756

Richard Nugent, Whig MP for Bristol, found himself apologizing to the Duke of Newcastle for the apathetic Whig support in Bristol. Nugent explained that at present the corporation was occupied with administering justice through the assizes (quarterly criminal courts) and feasting. He wrote to the Duke, 'Their mouths are full of turtle' and reassured him that as MP he had a suitably stirring address prepared. He drolly supposed that his corporate address would be well received, not least because it would be delivered along with turtle: 'Their address, I daresay flatter myself, will partake of their diet, for turtle is wont to inspire warm, kind, and vigorous sensations.' The turtle must have gone down well, because a few weeks later the Bristol Corporation expressed its support for the Duke's Whig government.[10]

London and Bristolian aldermen, Yorkshire gentry and other connoisseurs of turtle need never have gone without turtle. Turtles shipped from tropical waters and served on English dinner plates evince well the extent to which regular and sustained trade routes shaped elite consumer habits.

Buying a turtle from a tavern or turtle warehouse, however, was not the only way to secure this delicacy for dinner. It is clear from a number of letters and private registers that the donation of a turtle was a sociable gesture and a means of enhancing one's reputation. The dinner register of the Royal Society's dining club, maintained by its treasurer, reveals that turtle played an important role in socializing and the politics of dining. The dining club met regularly on Thursdays at the Mitre Tavern on Fleet Street and was also known as Thursday's Club. Fleet Street separated the City of London from the City of Westminster and came to represent the social divide characterized by the two broad sorts of turtle eaters: alderman and aristocrat.

In the 1750s a turtle was a highly prized treat, so much so that it was rewarded handsomely. The record of a meeting of Thursday's Club on 4 October 1750 documents the intentions of Andrew Mitchell Esq., MP for Aberdeenshire, to 'compliment the company with a turtle'. This particular turtle was at that very moment en route to England from the West Indies. The taste buds of the club must have indeed been tickled since the club voted that 'any gentleman giving a turtle annually should be considered an Honorary Member during the payment of that annuity.' The move was without precedent. It must have been with quite some disappointment to both Mitchell and the eager diners that a week later a calamity foiled all plans for a turtle dinner. The dinner register recorded that the courses has been changed, the turtle 'happening to die as the ship came up the Channel'. Proof indeed that one ought not to eat one's turtle before it is dressed.

James West, a secretary to the Treasury, was a more successful turtle donor. In 1755 he donated a 70-lb turtle on 7 August, followed by a 60-lb turtle on 18 September. In that same month Admiral Lord Anson had also given a turtle weighing 115 lb to the club, making that late summer's turtle trio a culinary high point for the gentlemen of the club. The turtle dinners in August and September had served 23 members and 18 members, respectively, filling stomachs and currying favour. The turtle had supplied four courses, and secured West honorary club membership. He renewed his membership with a turtle the following year too. Between 1768 and 1772 West was made president of the Royal Society and the dining club, suggesting that turtle did indeed travel far with the fellows of the Royal Society.[11]

Turtles could be used to cement social relations in other ways too. In 1766 Thomas Bradshaw Esq., a Lord Admiral and undersecretary to the Treasury, wrote to his friend George Selwyn with pleasing news:'I have heard by accident that you want a turtle for a respectable alderman of Gloucester, and I am happy that it is in my power to send you one in perfect health, and which I am assured by a very able turtle-eater appears to be full of eggs.' Bradshaw was able to use his connections and leverage to secure a turtle for Selwyn, who as MP for Gloucester was naturally eager to ingratiate himself with a prominent Gloucester alderman. On another occasion, Selwyn had been invited to dine at Spring Gardens, Ranelagh, on 26 July 1774. These gardens were a fashionable retreat near London and included landscaped grounds, private supper boxes and pavilions. He later wrote to his adopted daughter Maria, or 'Mie Mie', fuming that although 'we ate very good turtle' his anticipated social engagement with a close friend had been foiled; 'this damned turtle party has kept me so late I doubt if I shall see him tonight.' For Selwyn, at least, it was unthinkable that he would decline a dinner invitation when turtle was on the menu.[12]

Turtle was, though, not always received favourably, being an acquired taste. In the early 1750s Valentine Knightley, MP for Northamptonshire, hosted turtle dinners for the local gentry and clergy as a means of endearing himself to his constituents. News of this turtle surfeit reached Horace Walpole, aesthete and MP for Callington, who took it upon himself to write witheringly: 'Knightley has been entertaining all the parishes around with a turtle feast, which, so far from succeeding, has almost made him suspected for a Jew, as the country parsons have not yet learnt to wade into green fat.' Knightley's turtle was alienating him from the rural Anglican clergy, who were used to plainer fare. The 'green fat' was an indulgence too far, and as Walpole's anti-Semitic jeer indicates, became conflated with money and otherness. The taste of the turtle's 'green flesh' was anathema to the middling sort.[13]

In the eighteenth century, beef-eating John Bull came to represent *English* virtue and character. Historian Ben Rogers has shown how Englishness was strongly related to eating habits and contrasted with those of the French. For the middling-sort, beef or mutton was accompanied simply with gravy, mustard or plum pudding. French cuisine, with its veritable cornucopia of sauces, stews and soups, was seen as affected and artificial. Truly *English* food was plain and reflected political values such as simplicity, virtue and manliness. A simple roast beef or joint of lamb starkly contrasted with braised and stewed meats accompanied with fussy sauces or soups.[14]

The eating habits of the middling sort were not, however, those of the aristocracy, who dined in greater variety and plenitude. The material world of elite dining culture expanded rapidly in the eighteenth century, as new eating habits engendered new ways to eat:

porcelain tea-drinking paraphernalia, a greater range of cutlery and ever-increasingly ornate and diverse silverware.

Barrel-hilted cutlery may have been an exaggeration, but turtle eating did give rise to some specialized tableware – aside from the *London Magazine*'s alleged chisels and assorted instruments. Soup tureens on a grand scale, in the form of a turtle, were the height of extravagant silverware. Several extant examples from silversmiths like the London-based Paul de Lamerie produced in the 1750s are suggestive of the esteem held for the turtle soup within. Silverware on this scale was produced either for elite consumption or civic purposes. In 1750, John Hill, commissioner of customs, commissioned a tureen from de Lamerie. Hill's wealth and position made him no stranger to a turtle dinner; indeed it was under the auspices of the Board of Customs that turtle-laden ships paid duties.

An impressive silver turtle tureen might be assumed to have taken a central place in a dinner. However, soup was in fact the first course served in the eighteenth century. Soups or stews were first, followed by meat or fish dishes. Lastly sweetmeats, tarts and fruit were served. Soup was no perfunctory 'starter'; Georgian polite table etiquette demanded that the host would supervise the serving of the soup. Moreover, it was poor form for a guest to refuse soup, it being far preferable to make a show of taking a few spoonfuls than sit with an empty soup bowl. The serving of turtle soup at the beginning of the dinner, then, was an opportunity to distribute largesse and engage in sociability. To feed guests so conspicuously well, from the very first course onwards, made a turtle dinner a memorable occasion. Turtle also made an excellent 'remove' dish, a conversation starter and distraction between the first soup course and the second more substantial course. As servants removed the dishes and cutlery and replaced them to prepare for a second course, attention could be focused on impressive removes such as venison, beef, boar or a tureen of turtle and turtle fins. Diners were free to pick at will from the dishes that were placed closest to them, and determined diners would sometimes offer a small bribe or favour to footmen to impel them to arrange the table favourably. As a result, if turtle was a dish other than the mandatory soup serving, actually tasting a morsel of turtle was not guaranteed if it was out of reach at the head of the table. Would-be turtle diners would have been much obliged for the use of a barrel-hilted knife and fork. Some diners, though, would not have been so crestfallen. Turtle's flavour and reputation did not endear itself to individuals like Lady Anne Barnard, who confessed in her diary, 'I have lived in terror of having a turtle dinner all the days of my life.'[15]

Turtle travelling the turnpikes of England has revealed the surprisingly rich world of a controversial ingredient. Turtle was the delicacy of the few, criticized by the many. It was also, as a luxury good, a means to secure and enhance social relationships. The plentiful turtle satires and dramas articulate the extent to which turtle consumption was conceived of as a moral or social issue. Turtle was not a ubiquitous household staple. However, the next ingredient we shall look at, bear grease, was a necessity of elite life. In comparison to the turbulence created by the turtle, the everyday acceptance of bear grease may appear mundane, but it was a vital component of self-presentation in public for those who aspired to polite gentility.

CHAPTER 5

Bear Grease for your Powdered Wig

WAXED, STIFF MOUSTACHES WERE to the Victorian gentleman what a properly powdered wig had been to his Georgian counterpart: indispensible, respectable and masculine. Yet, by the 1850s, grooming and sartorial trends were sufficiently different from the Georgian period that contemporaries regarded those of their grandparents and forebears with baffled amusement and disdain. The bear grease that was combed as wax into a

2 *A Bear*, James Sowerby (1756–1822), graphite

moustache or used as a pomade to render one's hair stylishly sleek and slick in the nineteenth century was employed in the previous century in a very different manner. In the Georgian period bear grease was used principally to fix scented or coloured powders to wigs, as well as providing the necessary texture and adhesion to comb hair or wig attachments supported by cotton pads and gauze into the voluminous styles of the day. Powders served to conceal the natural hair colour of the wig. Bear grease was also a restorative pomade for the scalp or a necessity for soft, strong and luxuriant hair.

Wigs for both men and women, even if modest, were a hallmark of gentility and respectability for most of the eighteenth century; as such, those with pretences to gentility or desiring to appear au fait with the latest trends wore wigs. Certainly, in addition to the elite, the middling sort – the clergy, legal men, physicians and merchants – wore wigs in public life, and so too did some servants, labourers or tradesmen, though these were simpler in style, cheaper and less frequently or elaborately dressed. Wigs of course had an aesthetic function that appealed to Georgian cultural tastes, but they were also, to some extent, a pragmatic choice. Cropped or shaven hair, otherwise hidden under a wig, was easier to keep clean and lice-free. Likewise, wigs also disguised baldness, a state that was stigmatized by its association with syphilis or gonorrhoea (patchy hair loss being a symptom of the later stages of these venereal diseases). So too did the lead and mercury used in cosmetics or medical treatments wreak havoc on the scalp, in addition to causing sores and tooth loss. Wigs thus concealed the unfortunate vicissitudes of ageing and disease.

The trades of the barber and the 'peruke' maker or wigmaker were closely allied, and usually a wigmaker would shave and dress his customers. In addition to making and repairing wigs, a wigmaker would also delouse and recurl wigs. Like a hairdresser too, a wigmaker might also dress a wig with an array of scented and grey, white, blue, pink and violet powders, along with ribbons and other baubles. This trade, at least in London, was reliant on another: that of the 'hair merchant'. A hair merchant sourced hair and sorted it by colour and quality, pre-curled it and dressed it so that it could be bought wholesale by wigmakers in 'town', i.e. London, and more readily woven or sewn into silk and cotton pads. Provincial wigmakers had the onerous task of securing regular stock on their own. The diary of the early eighteenth-century Manchester wigmaker Edmund Harrold reveals, along with his passion for books, 'doing the wife' and drink, the precariousness of his trade. Harrold was born into the middling sort, his father was a comfortable tobacconist, but Harrold experienced a sort of downward mobility as his trade was not lucrative and based on credit, meaning that he was a perpetual debtor. His customers too frequently bought wigs on credit, paying Harrold in instalments. Significant time and expense was spent too on securing stocks of hair and the proper tools for his job. Peruke makers, to make a great deal of money, needed to cater for the elite and create wigs in the latest styles. Harrold, unlike some other provincial peruke makers, did not travel particularly far afield either to ply his trade or to remain *en courant* with the French styles craved by the rich. Instead, even in Manchester, competition was stiff and sometimes Harrold had to get by on a sluggish daily trickle of repairing and dressing wigs with the odd order for a new wig.[1]

In contrast, the trade could be a lucrative one with a degree of social cachet; London-based wigmakers and hair manufactories could do well, with some, such as Ross's Ornamental Hair and Perfumery Warehouse, receiving a royal warrant. Ross's wigs were said to add to their esteemed wearers 'a loveliness to youth and a respectability to age'. His advertisements boasted that his wigs were worn to 'every fashionable gala' by the aristocracy and gentry, and particularly 'those of taste and fashion', thus 'incontrovertibly proving' the pre-eminence of his wigs. His wigs were, at the low end, priced at two or three guineas. At the very top, prices anywhere between 10 and 100 guineas were usual. This was a huge sum, especially since Edmund Harrold had only commanded between 4 and 24 shillings for his wigs.

In addition to wigs, hair manufactories also, at least from the 1750s, sold bear grease. Bear grease was seen as a superior product to the more common tallow pomades, though in actuality both must have been rather unctuous and malodorous, especially in the summer – hence the necessity of scented powders and regular dressing by a wigmaker to delouse and freshen the wig up. Bear grease was not by any account cheap; small pots could sell from between one shilling and a guinea, or grease could be purchased for two shillings an ounce. A cheap barber could give a shave and dress a wig for sixpence. Bear grease was well out of the reaches of London's labouring poor, who might have otherwise spent a shilling on 4 lb of meat, 1 lb of butter, 3 oz of tea, 2 lb of sugar or 2 lb of cheese. As it was, much of London was malnourished or just managed a subsistence living. At this price it is not surprising that bear grease had many imitators, many finding an adequate substitute in the tallow of cattle or pigs. Even if used sparingly, bear grease would require regular repeat applications. The profusion of toiletries and cosmetics in the Georgian period was part of a large explosion in consumer goods. In the eighteenth century landed wealth increasingly gave way to wealth accumulated through trade, irrespective of blood and acreage. London's high bourgeoisie could live and dress like the landed gentry and aristocracy; indeed, the sons and daughters of trade married into these families, providing much-needed cash in exchange for a title that would efface, eventually, their roots. The interloper or individual hitting above their station was a source of cultural anxiety and criticism. With the right wig, clothes and manners, the lowborn might easily rub shoulders with the highborn; thus the beau monde took pains to display its distinction from and superiority over the nouveau riche. Georgian satirical caricatures are replete with cruel waspish depictions of ruddy country folk returning from London à la mode, artlessly aping the fashions of the metropolitan bon ton; snide parodies of the mercantile rendering the genteel vulgar; lumpen maidens in fine muslins, or prostitutes plying their trade in chintz cotton gowns. The beau monde was a significant arbiter of taste, even in a society with an increasingly substantial 'polite and commercial' middling sort. Bear grease was a marker of distinction and quality even if it was open to imitation; mere tallow or lard could be distinguished as such by those in the know. In the same way that London's best homes would suffer to burn only beeswax or spermaceti candles and not cheap tallow candles, so did the most discerning wish to dress their wigs or lacquer their locks with bear grease.[2]

In *Sketches of Animal Life* (1855) the parson-naturalist John George Wood supposed that 'ninety nine of every hundred pots of bear's grease are obtained from the pig' since, he assumed, bears are not quite plentiful or so easily killed as to supply the quantity of grease that was consumed annually 'even in this metropolis', let alone the whole of England. Wood preferred to refer to bear grease as 'lard perfume'. His healthy sense of scepticism was perhaps a little misplaced; after all, a lot of bear fat was imported into Britain in the nineteenth century for processing. In the United States, thousands of barrels of bear grease were shipped to New Orleans for both export and domestic consumption. Arkansas, the 'Bear State', supplied much of this bear grease; over the course of the nineteenth century the bear population there decreased from 50,000 to a few hundred. Another writer in the 1850s claimed, that, though 'Bruin has been lately declared a humbug', some of it was indeed genuine. Admittedly the public grumbled that bear grease did not flow quite as freely as barbers would have them believe, and charlatans no doubt hoped to sell pig lard as bear grease but, this notwithstanding, it is certain that a significant quantity was genuine.

In eighteenth-century London, the bear grease on sale came from American and Russian bears. These bears were often imported and fattened up on the premises of London hairdressers, barbers, perfumers and wigmakers. Every winter barbers and hairdressers slaughtered around 50 bears in London. The 'shaggy denizens' of the Russian forests, 'regular passengers between St. Petersburg and London', yielded up their precious fat to smooth the fair locks of the rich.[3]

Reeve's Perfumery Warehouse and Ornamental Hair Manufactory on Holborn Hill killed a bear a month throughout the 1790s to satisfy demand. Likewise, Lewis Hendrie Perfumer and Vickery's Perfumery Warehouse frequently advertised the imminent slaughter of a fattened bear. In Dublin too, Jones's Perukemakers and Hair-dressers supplied bear grease to an image-conscious elite. Lewis Hendrie, in the 1780s, became so well known for his bear grease that he was the inspiration of the satirical poem *An Ode to Mr. Lewis Hendrie, Principal Bear Killer in the Metropolis of England, and Comb-Maker in Ordinary to His Majesty* (1783). A literary reviewer supposed that some 'wag' had wanted a laugh at the expense of a 'poor devil of a perfumer'; however, one should not feel too sorry for Hendrie – after all, he would 'sell more pounds of bear grease, than the bookseller will copies of the poem'. Hendrie really wasn't such a 'poor devil'; he was also 'Perfumer in Ordinary to the Princess Royal' so no doubt turned a respectable trade.[4]

In the wild bears fatten themselves up on vast quantities of nuts, acorns and berries prior to hibernation; in captivity they were fattened on the premises with scraps and meat, though in leaner times they lived only on bread. Before slaughtering them, proprietors would advertise the weight of their bears in order to draw interest and boast of the grease they would produce; some bears were between 400 lb and 700 lb. Moreover, proprietors crowed about the costs incurred in fattening the bear, done at their own expense. These bears were usually kept in yards or in cages. After slaughter the bear would be hung on chains, skinned and then fat was removed from the flesh, to be boiled for several hours to produce thick grease. Customers were invited to witness the removal of the fat in person and guarantee the veracity of the grease, or they could send their servant in lieu. A barber in the vicinity of St Giles's Church hung up a bear on chains outside the second storey window of his premises, putting up a placard announcing that customers might come and cut off the fat themselves, providing they brought along their own gallipots, small earthenware pots used for ointment by apothecaries.

Writing in the Victorian era, John George Wood doubted that the bear grease in Georgian London had indeed been anything other than pig fat. He believed that perfumers and hairdressers had kept bears on the premises to 'poke' when customers were around, drawing in a crowd to gawp at a bear and show that they had the real thing. But the bear carcass was actually a 'nice fat large hog' wrapped in fur – a pig in bear's clothing. This ersatz bear hung in the window, so he claimed, with people paying to 'purchase permission to rub their heads into the bear to ensure getting real bear's grease'. This tale is almost certainly apocryphal; high-society ladies and gentlemen were not likely to stick their heads inside a bear in public. The sending of a servant to observe the cutting of the fat was almost certainly a step taken to ensure the authenticity of this expensive consumer good. At close quarters a pig hanging in a furry blanket was rather different from a fattened bear. In any case, the fat of a pig had different qualities from bear grease. And one could, apparently, tell the difference.

There were two sorts of bear grease. One had the consistency of thick olive oil, softer than pig fat, and was procured by boiling the intestines and surrounding fat; the other was like 'frozen honey' and made from the kidney and associated fats. When fresh both types of fat were said to 'stink intolerably' but the barber could produce less odorous grease if he made certain to hang up the fat and clean it prior to boiling. The grease could be perfumed later by the barber or at home. Certainly those purchasers sending their servants to acquire the raw grease would have recognized the distinctive smell. This could be masked a little with the pleasant scents of rose or lavender oil. The New London Toilet (1778) suggested taking 3 oz of bear grease, boiling it with 1 pint of red wine and then adding some camphor-scented Artemisia. Alternatively honey or sweet almond oil would give a better scent. Rubbing a little of this pomatum over the head every morning was supposed to quicken the growth of the hair. The fat of a pig had a different scent to bear fat, although rancid lard was sometimes passed off as bear grease if it had acquired through age the sufficient 'perfume'.[5]

James Rennie, in his The Art of Preserving the Hair (1826), marvelled at how 'this disgusting stuff ever became popular'. He supposed that the use of bear grease was due to some 'far-fetched analogy' of bears being rather hairy, and the product becoming acceptable amongst 'decent company' thus earning a place in the 'fashionable toilet'. If people did have genuine bear grease in their possession and insisted upon using it, 'though they should smell like a bear', they should weaken it a little with beef marrow; this 'adulteration' would temper the oiliness of the bear grease.[6]

Not all the bears that were imported into Georgian Britain were necessarily intended for the grease trade. Bear baiting, though less popular than it had been in the seventeenth century, still persisted, and dancing or performing bears walked the streets. However, bears that were here for these entertainment purposes could end up boiled and potted in gallipots. In 1805 Mr Bradbury, a clown at the Royal Circus near Blackfriars Bridge, bought a young bear in Liverpool and brought it to London on the roof of a stagecoach. His bear was tame enough to walk around without a muzzle and was taken by Bradbury to London taverns and parties to earn money. Here the bear, with a hat on his head, would sit among company and take his bread and beer. But the duo separated in early 1806 when Bradbury's services as a clown were required in Manchester for several weeks, with the bear left behind in London. Left in the charge of a neglectful porter, the bear was kept in a brick shed in a yard, chained to a post and fed intermittently on bread. The sweet bear's docile temper frayed, and in a rage fuelled by hunger and confinement he broke loose during the night. The bear that had tamely followed Mr Bradbury around like a dog became a fearsome fugitive. The roaring of the bear and the furious barks of the circus dogs brought the circus's carpenter running. The carpenter was in turn pursued by the bear and escaped only by running up a stone staircase, his coat tails torn off. The 13-year old son of the circus's billiard room proprietor was drawn to the noise and found himself face to face with an angry bear. The bear quickly chased him down and brought him to the ground with his two forepaws. The boy landed on his face, and the bear tore off the back of his head; a crowd pulling on the errant bear's chain and poking him with pitchforks managed to separate the two. The boy, 'in a gore of blood', was sent to Guy's Hospital for surgery. The bear then incurred the wrath of the crowd. It was shot, had its throat cut, or its head split in two by a soldier with a spade, depending on

3 *Study of a Bear*, Sawrey Gilpin (1733–1807), graphite

the newspaper or periodical. When Mr Bradbury returned from Manchester he apparently found that 'all that was left of the bear was the shaggy skin and a large supply of pots of bear's grease in a neighbouring hairdresser's window'. The circus had evidently sought to profit out of even the most unfortunate of events; the bear that had torn off the back of somebody's head was now worn, in a manner of speaking, on the head of appearance-conscious Londoners.[7]

An Englishman called Charles Edwards, on his return from the Grand Tour, supposedly wrote a series of 'letters' in the late-Georgian periodical *Blackwood's Magazine*. He assumed the bearing of a man who could now look at England through the eyes of a foreigner, though naturally his heart remained resolutely that of an Englishman. These letters aimed to cast an eye on the foibles of the English and convey a spirit of the times. In 1825 Edwards recalled the case of 'two rogues', hairdressers who had become somewhat notorious in Regency London. He called this case a 'laughable experiment' on the 'force of truth or puff'. A hairdresser living in the vicinity of the

Exchange was thriving on a steady trade in bear grease, a state of affairs that did not escape the notice of a neighbouring hairdresser. This man took to using an 'inexpensive unguent', probably lard or tallow, instead of bear grease and potting it up as the genuine article. The 'true dealer' who, at least according to Edwards, kept 40 live bears in his cellar, soon got wind of the 'imposter'. That the man had 40 bears is almost certainly the exaggeration of a Georgian urban legend; four bears might easily turn into 40 as tongues wagged on the grapevine. The hair manufactories and barbers did sometimes advertise that they had several bears on their premises, so three or four bears in a cellar is not improbable. To assure his customers of the genuine nature of his bear grease this 'true dealer' killed a fresh bear and hung it in the window and stuck a placard below it: 'A fresh bear killed this day'. The 'imposter' peddling his fraudulent bear grease had but 'one bear in all the world' and, so it was said, in an act of audacious and inspired deception sneaked the bear out under the cover of darkness only to bring it back in the morning in full sight, to make it look as if a different bear was arriving to be slaughtered and potted up. The fraudster wrote in his window, 'Our fresh bear will be killed tomorrow'. His rival, not to be outdone, claimed his nemesis to be a vile imposter and invited patrons to see their bear grease 'with their own eyes, cut and weighed from the animal'.

Some three days passed, and the shrewd bear grease counterfeiter came back with a rather brilliant advertising scheme that sought to turn his proverbial sow's ear into a silk purse. To show neither hide nor hair of a bear might otherwise have been an embarrassment but virtue was made out of a necessity. After all, so the rogue's new placard cheekily announced, the best physicians had concluded that the grease 'obtained from a bear in a tame or domestic state' would never make anybody's hair grow. Instead, he had 'formed an establishment' in Russia for catching bears in the wild, cutting off the fat, and potting it for 'London consumption'. There was then no need for a bear in a cellar, a skin in the window or a carcass hanging. The hairdresser had written all over his premises 'Licensed by the Imperial Government'. With the airs and graces bestowed by a fraudulent Russian seal of approval, he had carried the day and severely antagonized the purveyor of genuine bear grease. Charles Edwards, at the time of writing his 'letter', informed his readers that this dealer was making an application to the Court of Chancery to seek redress. This 'true dealer' was, however, not unfamiliar with the legal proceedings, especially those of the City of London – he had appeared before an alderman several times as a 'nuisance'.[8]

The nature of the sort of nuisance that hairdressers could cause the authorities is recalled in the memoirs of the lawyer Cyrus Jay, a legal man who had frequented a tavern in Fleet Street for some 55 years. To illustrate the want of professional and trade knowledge that he found in some aldermen, Jay recalled an incident from the early 1820s, around the time that Edwards had written about the case of 'truth or puff'. The Lord Mayor, who was at the same time an alderman of the City of London, whilst incumbent at Mansion House heard an astonishing tale fed to him by a hairdresser's accomplice. About 'the time when bear grease was in such demand' a woman appeared before the Lord Mayor and told him that while admiring the wigs displayed in the window of a hairdresser's she had felt something grabbing her ankle. Looking down in horror she saw that it was the paw of a bear; 'My lord I nearly fainted away and I am now in such a state that I shake like a calf's foot jelly'. The Lord Mayor requested the hairdresser to come at once and he heard him apologize profusely for the unfortunate incident. The hairdresser explained that he had 33 bears he was going to slaughter since the demand for grease was so high, but in future he promised to slaughter them in the country and not the city. He would as a gentleman also ensure that the woman was conveyed by carriage at his own expense to her home in Hammersmith. According to Jay, both had duped the Lord Mayor: the lady was the hairdresser's sister, and the hairdresser had arranged the whole affair as a publicity stunt. This case was all puff; the hairdresser 'naturally had only one bear, not thirty-three', and once news spread about the hairdresser with 33 bears he did a roaring trade selling not only bear grease but also a 'great deal of other animal's grease'. The surfeit of 'bear grease' was deftly explained away by the 33 phantom bears in his cellar. Jay's point was well made; perhaps aldermen ought to have a little more trade knowledge. After all, only the dim-witted or credulous would fall for such a plot. A mere passing knowledge of the trade of hairdressers, barbers and wigmakers would have enabled his lordship to smell a rat. There was no such abundance of bears in London cellars. The deception came to the attention of other aldermen who, according to Jay, would not allow the Lord Mayor to hear the end of it.[9]

It is almost certain that the rival hairdressers described by Edwards were no other than Mr Money and Mr Macalpine, both operating out of premises on Threadneedle Street. Though it is uncertain which man was the purveyor of truth or puff, Macalpine, of 48 Threadneedle Street, proudly advertised himself as haircutter and peruke maker to George IV. In 1825 both men appeared at Mansion House to account for the nuisance their bears were causing the public. Complaints had been made about the roars of the bears, and the crowds the 'bears attracted through the doors, and blocked up the thoroughfare'. Mr Money's bear could quite readily stick out his paws and seize passersby. Macalpine's bear was almost entirely at liberty and so might, if it so pleased him, vent his wrath on 'any of His Majesty's subjects that come near him'. Moreover, with the bears in want of food or the solace of fellow bear society, the barbers' premises resounded with 'hideous howls' at night. Mr Money's assistant hairdresser, 'his hair cut and curled with mathematical precision', was described by the Lord Mayor as a 'spruce young barber', and this young fellow sought to defend his employer against accusations of neglect. The bear cage was not near enough to the window to cause accident or injury to the public. Bear paws and claws could not reach that far. The public could only see the bear, and surely there was not any harm in the mere sight of a bear. The Mayor disagreed and ordered it removed from display in the window. On the matter of Mr Macalpine's bear, the Mayor was also disapproving. Macalpine insisted that his bear was as harmless as a lamb and was now chained in his cellar. Moreover, having already killed one bear to provide grease for his business, he would not be persuaded so easily to kill his amiable young bear to gratify his lordship. The Mayor told the hairdresser that he could do as he pleased in that respect, as long as the bear was not a nuisance to others or remained loose.[10]

Rival bear grease dealers were remembered for decades after. In 1868, the *Gentleman's Magazine* wrote of how the long history of wigmakers in the city persisted into living memory: 'they kept shop here until our own time for we remember rival dealers in bear's grease, *ecce signum* the bears skins at the door.' *Ecce signum*: behold the sign – skins hanging as proof. It was not only the barbers and bear grease of Victorian London that inspired Charles Dickens, he also drew heavily on their Georgian counterparts. Mr Money and Mr Macalpine had their literary manifestations in Dickens's work. Mr Jinkinson was a long-deceased and fondly remembered barber in *Master Humphrey's Clock* (1840). The epitome of the Georgian master barber, 'cutting and curling was his pride and glory', and all 'his money was spent in bears'. Jinkinson's bears growled and gnashed their teeth in the cellar, and his first-floor window was decorated with a grim array of bear heads. In a final flourish, Dickens's anecdote poked fun at the barber's fond attachment to his bears. When the ageing Jinkinson was confined to his bed in low spirits, the doctor gave his bears a poke. Their roars roused Jinkinson to a cry of, 'There's the bears!' His spirits and 'pride in his profession' were restored.

The advertising and bear gimmicks helped barbers sell their grease. But they also attracted criticism as tricksters who were fuelling a trade in pernicious cosmetics and perfumes. A letter to the editor of the *Statesman*, titled 'Perfumes and Razors', condemned the 'loquacious puffing' of hairdressers and perfumers. The perfumer who boasted 'superior fragrances' stank like a polecat, or the drunkard with his 'mulberry face of habitual intemperance' promised his customers a blemish-free visage. Worse, though, were the advertisements for bear grease, 'matchless for promoting hair growth', words that sprang forth from a bald barber whose skull had long been stripped of its fine locks.[11]

In January 1763 a gentleman wrote in twice to the *St James's Chronicle* to expose 'filthy fashions'. One of his peeves was the stench of the dirt trodden into the hems of society ladies' negligees, morning gowns or trailing loose garments that were being worn outside. The dirt must have been animal and human, hiked in from the street or walks in the park. Said gentleman had recently had the 'happiness to drink tea with a dozen of the most polite females in this metropolis' but the 'disagreeable smell' of the company offended his 'olfactory senses'. Determined to expose the source of the unpleasant odour, he took to turning up the ladies' 'dirty tails', an impish act he assured the editor and readers, 'all done in good humour'. Their graceful fashion, he concluded, was in truth a very dirty one; these dresses ought to sweep drawing-room carpets and not the streets.[12]

The main source of ire for this fashion-policing gentleman was, however, these ladies' bear-greased and 'frizzlated' hair; he reserved an entire letter to vent his spleen. During the war with France, a certain French fashion was creeping through 'this metropolis' that 'must infallibly infect the whole nation'. He hated this fashion not purely for its Gallic origins; it was a 'filthy one', and the French, 'with all their politeness are, in some respects, a nasty people'. The infectious fashion which this gentleman had in mind was the trend for big hairstyles among ladies – 'an additional growth of hair' when they appeared in public. Voluminous hair was not yet, among women, necessarily created with a wig; women's big wigs would become especially popular in the 1770s and 1780s. In 1763 it was 'frizzlation' and bear grease that gave hair an extra lift. The hairdresser, taking strands of hair between his thumb and forefinger, would wrap them in squares of paper and then burn them with a hot iron; 'her ladyship now looks like a sunflower.' After removing the papers the hairdresser would daub on 'at least half a pound of grease' and then powder the hair with a pound of flour. This look, according to the writer, resembled the stuffing of a chair or a 'state of confusion'. Worse still was the odour this 'pudding' produced: bear grease and tallow in burnt hair, a pudding that was set for about three months. In this period it was not in 'a lady's power to comb her head' and the result was not pleasant; 'Her ladyship stinketh.' If such women wanted to either find or keep a husband they would do well to desist in producing such a 'tainted breeze'. This curmudgeonly gentleman doubtlessly seethed for years, as the trend for greased hair did not wane for decades. But he was not alone in his censure; Georgian periodicals and caricatures are replete with criticism and satire. Exotic animals sometimes served to underscore the follies of feminine fashion. *Sleight of Hand by a Monkey* (1776) depicts a cheeky monkey perched on a wall, snatching the wig off a passing woman, thus revealing her plucked bald head. Likewise *The Feathered Fair in a Fright* (1779) features two bewigged young ladies fleeing from some cross-looking bare-bottomed ostriches. The ladies' wigs are crowned with ostrich plumes, and the birds look set on avenging their loss.

Sartorial winds of change in the 1790s, especially among the young, made perukes increasingly redundant, though these sorts of wigs were still worn by more conservative individuals into the early nineteenth century. A powdered wig, on both men and women, characterizes for many the Georgian age, but for the last few decades of this era it is instead the 'Brutus', the 'Caesar' or other hairstyles that took their cue from classical antiquity that capture best the spirit of the age. Men's hair was cropped short and parted with wax, or styled as a dishevelled mass of curls, artfully arranged and set with bear grease oil or pomatum. Likewise women's hair was loosely arranged in a chignon, curled into ringlets or curls, helped along by a little bear grease to fix them in place. The end of the peruke was far from the end of the trade in bear grease.

CHAPTER 6

'The Product of the Civet's Posteriors'

'THERE HAS BEEN A time when the product of the civet's posteriors was in the highest estimation with the ladies and effeminate men.' So the teacher-turned-clergyman William Fordyce Mavor hoped to instruct and entertain the young readers of his *Natural History for the Use of Schools* (1800) on the folly of civet, the waxy yellow-brown secretion from the anal glands of the civet cat, one of around a dozen small slender carnivorous mammal species from tropical Africa and Asia. The civet marks its territory with a pheromone called 'civetone'. The strong musky odour of civet hung in the air for decades as both men and women of means and fashion wore perfumes, used wig powders, and took snuff that was made with civet. Mavor continued squeamishly: 'the very idea of borrowing from such a source is not a little offensive to a delicate mind.' Clearly, the preceding generations of Georgians had been made of sterner stuff.

Civet-scented snuffs and fragrances epitomized luxury and decadence, depending on one's particular taste or position, for much of the eighteenth century. As with all fashions, which by their nature wax and wane, that which was once loved is eventually scorned, and so it was with quite some relief that late Georgians noted the passing of civet. The musky base of civet perfume that complemented so well the natural, and infrequently bathed, body of its wearer became much maligned and instead lighter floral-based scents were the preferred late eighteenth-century *l'air du temps*. This change was not only a generational change in aesthetics and preferred taste, nor was it solely prompted by a sense of revulsion at the unavoidable and frankly disagreeable fact that civet, as a product, came from a civet cat's bottom. Instead it was the strong association of civet with indulgent luxury and corruptive effeminacy that marked the watershed in the use of civet. The trade had been lucrative, and civet, especially unadulterated English civet, was perhaps one of the most expensive consumable consumer goods of the Georgian period – even more so than bear grease or turtle.[1]

Elizabethan Londoners were familiar with civet and associated the odour with a certain kind of sexual musky sensuality; civet was both an aphrodisiac and a perfume. Shakespeare's King Lear had lustfully cried, 'Give me an ounce of civet, good apothecary, to sweeten my imagination', a pearl that more literary-inclined Georgians knew well and delighted in. Much of the civet used by apothecaries and perfumers in the sixteenth and seventeenth centuries came from Cairo, Calcutta and Basra as well as from Ethiopia. Merchants, often with many hundreds of civets, would sell their product far and wide. It was thought that a warm climate produced the best civet and according to eighteenth-century apothecaries, perfumers and naturalists this foreign civet would be the best were it not for its adulteration; it was alleged that the non-European traders tainted pure civet with odorous plants, drugs and even honey or the pulp of raisins. Instead, for the European market, it was Dutch civet that was preferred because it was believed to be purer. Civet merchants in Amsterdam imported their civets and reared them to harvest their scent. The Dutch put special certificates on their sealed pots of civet to guarantee their authenticity and purity. Genuine civet could also be determined with varying degrees of accuracy by rubbing it into paper until the substance was absorbed without leaving a taint or residue. True civet was also neither soft nor hard; the consistency of butter or honey. Melting civet would also reveal any telltale lard and butter. Although the Dutch were the centre of European civet production, they were not necessarily the best. The French master apothecary Pierre Pomet, in his *Complete History of Drugs* (1684), said that the best civet was made in England despite the greater volume being made in Amsterdam. The London perfumer Charles Lillie thought that adulterated civet had not 'even half the smell or perfume' of the purer English or Dutch civet. Moreover purity was a matter of degree and even the better civet from Holland was 'so adulterated, and that in so great a degree, that true English civet fetches 40 shillings per ounce'. Foreign civet usually sold for less than half that price. By Lillie's reckoning, in order to produce a desired fragrance a perfumer ought to use twice the prescribed amount of civet, on account that civet was commonly diluted with cheaper substances. This prudence notwithstanding, perfumers are likely to have somewhat over-egged the pudding and at times produced suffocating concoctions. Civet imported into Britain also carried a heavy import duty; in 1777 the duty was eight shillings an ounce. The substance could be measured out and sold in minuscule quantities: a grain of civet sold for two pence. A perfume might contain between 20 and 40 grains of civet. For the poorest Georgian Londoners, two pence would buy coal for the day or a hot meal – about the price of a single grain of civet used to scent linen or a day's worth of eau de toilette. The high demand for civet was matched by steep prices.[2]

During the seventeenth century, and well into the eighteenth century, civet had many medicinal uses; its hot anodyne qualities made it an important treatment in Galenic medicine. Civet could warm the stomach, revive the spirits and relieve colic, as well as proving useful in treating sexual and 'hysterical' disorders. For gentlemen, an early eighteenth-century treatment for erectile dysfunction used 25 grains of civet. On the other hand, women who had taken too much pleasure in themselves ought to desist lest they spoil their uterus and complexion; civet taken in the form of a 'refreshing balsam' might aid such women who had 'debilitated themselves greatly'. Civet could be taken as a tincture, balsam or pessary and applied onto the intimate areas. By the mid-eighteenth century these medicinal uses for civet had by and large disappeared and instead civet remained in use only in perfumery. It was used to raise perfumes to a greater height, hide disagreeable smells in soaps, scent gloves and handkerchiefs, as well as in powders and snuffs. The scent of civet tended to linger and this was a trait that was put to good use in producing scented papers and boxes. Likewise, for better or worse, fabrics that came into contract with civet-based perfumes or cosmetics reeked of musky civet for months or even years after.[3] Woe betide someone who spilt it on the carpet, bedclothes or upholstery.

Charles Lillie was London's pre-eminent early Georgian perfumer and had his premises on the Strand near the Beaufort Buildings. He had set up shop around 1710 and was there until at least the 1730s; his handbill featured a rather doleful civet cat. Indeed the civet cat featured in the names or on the trade cards of a greater part of London's perfumers, with most trading from 'the sign of the civet', a wooden or stone civet cat marking the premises as a perfumery. The Old Civet, the Civet Cat and Roses, and the Civet Cat and Perfume Shop were just a few of Georgian London's perfume shops. Lillie was particularly renowned for his orangeflower- and violet-scented civet perfume as well as popularizing the use of snuff by London's beaus. Snuff was not particularly common in London prior to the 1700s. Taking snuff tended to be a habit of those who had lived or travelled abroad, though the habit soon became entrenched. Lillie gave classes to those young mercantile gentlemen who wished to become better acquainted with the 'ceremony of the snuff box' and apprise themselves of the 'most fashionable airs and motions' in which snuff might be taken. Snuff ought to be presented differently to one's friend, mistress or a stranger, and a certain artful flick of the thumb might indicate amiability or contempt. Lillie sold 15 varieties of snuff and counselled on how to add civet to snuff by rubbing it in one's hand to bruise the tobacco and work in the civet before returning the mixture to its box. The resulting snuff must have been pungent indeed. Fashionable mid-eighteenth-century Londoners with their civet scents and snuffs – though aware, on some level, of the less than savoury origins of their little indulgence

– might have paused for thought if they had had to obtain it for themselves with their own bare hands. This was unpleasant indeed, for the scent of the civet was born of a gruesome procedure.[4]

Civet cats were housed in narrow hutches, coops or cages to prevent them turning and biting the person sticking a wooden or silver spoon into their anal sac. The room in which the civet or civets were kept had a fire burning to maintain a muggy heat; good-quality civet came from cats that were kept warm and well fed. In a closed hot room the perfume of agitated civets was 'so copiously diffused' that few could suffer to be shut up with one for long. Civet cats fed on fish, eggs, millet, rice, milk sweetened with honey, and boiled flesh were thought to produce the best-quality civet; 'this odorous liquor is always in proportion to the quality and goodness of the animal's food'. Lillie supposed that the civets announced their desire to be 'civited' by rubbing their tails against the grilles of their cages; the eagerness to be milked arising from the pain caused by the scent collected in their glands. The following procedure is not, however, something any civet could have looked forward to with eager anticipation. First the civet-gatherer would pull the civet's long tail through an opening at the back of the narrow cage and then secure the hind legs. An instrument, similar to a long spoon used to scoop meat marrow from a bone, was then used to scoop and scrape the yellowish greasy civet out of the perianal glands. The freshly farmed civet was said to be so pungent as to induce giddiness, headache and nausea but apparently it became more agreeable through time as it aged. Male civets were preferred because females urinated into their 'civet bag', or glands, which spoiled the product. Some civet merchants would have their cats 'civited' more or less often than others. Civet merchants, apothecaries and perfumers during the height of the demand for civet calculated the highest frequency of 'civiting' possible before death. Some civets would die before they had yielded even ½ oz, especially if extraction was attempted more than twice a week or carried out by rough hands with a crude spoon. If daily extractions were attempted the civet would rarely live longer than a fortnight. The frequency with which cats were 'civited' also varied seasonally: every other day in the summer and twice a week in the winter. Pierre Pomet, who had been given a civet in 1688, struggled to get his civet cat to produce civet in substantial quantities; after 'some months' he had gathered only 1½ oz. He had tried to gather it daily and the cat was usually in an unproductive state of 'pain or apprehension'. Nothing was wasted in the production of civet. Bits of cotton were used to wipe the grates of the civet hutches as well as to clean instruments. This civet-scent infused cotton was known as 'civet cotton' and was sold for two shillings an ounce for the purpose of making sweet bags, perfumed sachets that were used to disguise unpleasant or embarrassing odours.

The value of civet fluctuated and in good times civet merchants and perfumers in London could reasonably expect to import Dutch civet, even with impurities and the addition of a duty, at an acceptable cost. However, when the price was high and there was commensurate demand for the product, this was when 'civiting' on a small scale sporadically emerged in London. The late Georgian editor of Charles Lillie's *The British Perfumer*, posthumously published in 1822, told his readers that 'when the use of civet ran high in London' at least 'several persons in town kept cats'. However the trade had eventually died: 'none are ever heard of now.' Likewise Lillie himself noted that civet could be acquired by English perfumers on a sporadic basis in London from those who 'keep the cats for pleasure and amusement' as well as from those who 'keep such animals for sale'. On an ad hoc basis, then, London perfumers in the eighteenth century were making the best of a buoyant market for civet by sourcing it in London. Civet cats were not particularly difficult to obtain even if it was difficult to make a successful large-scale enterprise out of making civet.[5]

English sailors off the coast of Guinea, for example, could pick up young civets for as little as eight or nine shillings or pay in Dutch coins. Thomas Phillips was captain of the *Hannibal*, a now infamous slave ship belonging to the Royal Africa Company. In addition to his terrible human cargo, Phillips brought aboard two civet cats that he kept in wooden coops and fed on boiled water and flour. They were far from pleasant to smell; 'the civetty scent they so strongly emitted was so offensive to me that I never cared to come near them.' As well as hundreds of slaves and the two civets the *Hannibal* was also carrying monkeys, baboons and parrots. When sold in England, cats like Phillips's were valued between £4 and £8 sterling each – a fine profit indeed. The Dutch enthusiastically imported the civet cat in large numbers to Amsterdam for the production and distribution of civet, leveraging on their mercantile empire constituted by the Dutch East India Company. The raising of civet cats to be 'civited' was said by foreigners to 'afford a considerable branch of commerce at Amsterdam'. English dictionaries, encyclopedias and popular natural histories that mention civets usually refer to the Dutch predominence in the civet trade and that the living cats were imported in 'great numbers'. Such large-scale operations were probably unusual in London, though civet merchant John Barksdale had some 70 cats. At Newington Green the 'civet house' was warmed by stoves to aid the production of civet, and contained a range of troughs and cisterns for feeding or watering the cats. Daniel Defoe of *Robinson Crusoe* and *Moll Flanders* fame, purchased his 70 civets for £850 in April 1692. This was not a successful venture for Defoe, mostly because he did not purchase in a straightforward and strictly legal fashion. Moreover, at the time of purchase Defoe was already knee-deep in debt. He actually paid £250 as well as some promissory notes and in exchange he was allowed use of the cats without ownership. At this point he could have made money by having them 'civited' to supply the perfumers of London with their sought-after civet. However, the promissory notes were worthless, so the original merchant John Barksdale sold the cats to a new owner instead; in the meantime Defoe was lumped with the care of the civets and associated costs until 1693. Worse still was that in addition to using his mother-in-law's servants to feed and care for his civets, Defoe also cheated her in the financial murkiness of his business practices. As a debtor harangued by his creditors, Defoe submitted himself to London's Fleet Prison in October 1692; the sweet smell of success eluded him and instead the musky odorous civet had sent him to gaol.[6]

Civet was often described as more 'grateful', or pleasant, than musk. Though civet was 'fragrant' it did need to be diluted as the smell of civet in its purest form was pungent and disagreeable. Its taste, God forbid, was described as 'subacrid' or 'bitterish'. Civet actually smelt better at a distance too, meaning that it was quite possible to walk into a room to be greeted by a 'grateful scent', only to sit down and find that the lady or gentleman sitting directly next to oneself, under one's nose as it were, 'stinks like a civet'. Sometimes a civety odour could induce 'swooning fits, convulsive motions, and suffocations'. This effect was compared, in unflattering terms, to that of horse dung and other excrements. One of the interests of seventeenth-century natural philosopher Robert Boyle was the 'mechanical production of odours', particularly those of animal excrements and sweats. Boyle told his readers an anecdote about a mathematician friend who was walking in Lincoln's Inn Fields. Passing by a dunghill, in the heat of summer, he detected a 'very strong smell of musk', but as he 'came much nearer to the dunghill that pleasing smell was succeeded by a scent proper to such a heap of excrements'. The sweet hay-like smell of horse dung from afar was unbearable in close quarters. Boyle drew a parallel between a dunghill and civet, drawing on his personal experience with civet cats: 'when I had been near the cages where many of them were kept together or any great vessel full of civet the smell was rather rank and offensive', 'but when I removed to a convenient distance the steams being less crowded and farther from the fountain presented themselves as perfume'. Georgian readers of Boyle's work on stinks and smells knew what he meant; they had smelt both civet and manure in abundance.[7]

The poet Alexander Pope mocked civet-scented courtesans in his *Satires* (1738): 'And all your courtly civet cats can vent,/Perfume to you, to me is excrement.' A few years earlier the rather less poetic and somewhat earthier verse *A Sign of the Times* (1733) supposed that were there no fops and beaus to dress, the 'poor civet cat would sh**e in vain'. The stink of civet, which was to some foul rather than fragrant, became an excellent medium for satire and social commentary. The use of civet certainly provoked strong feelings in William Cowper, who vented his spleen in the poem *Conversation* (1787): 'I cannot talk with civet in the room, / A fine puss-gentleman that's all perfume. / The sight's enough – no need to smell a beau.' It was not only the scent of the civet that was unappealing, but rather the effeminate wearer too. Men who cared too much about their appearance and indulged in feminine fashions or habits, fashionable though they were, attracted a great deal of criticism in Georgian print. Some commentaries pulled few punches; indeed, at the height of the trend for civet in the first half of the eighteenth century a particularly robust criticism appeared in a polemic published in the *London Magazine* in 1749 entitled 'Luxury Pernicious to Persons and States'. Luxury 'is the effeminate debaser of the soul' and from that 'pernicious root' sprang all manner of evil or sickness. Luxury and fashion created want, even in the midst of plenty. The 'charming country maid' was corrupted by 'swollen luxury' and perversely transformed into a 'tainted harlot of the town'. Like a fruit that was rotten even before it ripened, the former pastoral beauty showed her age in her 'withered visage and tormented heart'. In Georgian cultural imagination the very city of London and its attendant luxuries or temptations could be feared as a corrupting force in the moral fabric of the people. After attacking luxury the writer turned to the effeminate gentleman of fashion in a shrill but impressive piece of writing:[8]

Behold the gewgaw [gaudy and useless] butterfly, the beau, who looks like a girl, and smells like a civet cat, whose very words are female, and his gesture of the doubtful gender, who plumes himself upon his tailor's art, and, like a peacock, proudly spreads his gaudy feathers, whose utmost knowledge is the newest mode, and highest ambition the most admired dress; this pretty painted paltry creature is like a rich purse that has no money in it, or a foolish book, finely gilt and covered; the life of this poor Narcissus, like a transforming insect, entirely depends upon the cut and colour of his clothes; he lives but while they last, and when a fresher fashion or finer coat appears, he dies.

The foul smell of the civet was just another indication that all was not well in the kingdom.

Oliver Goldsmith's *A History of the Earth and Animated Nature* (1774) had little complimentary to say on the use of the civet. Goldsmith noted instead that in the realm of medicine 'at present it is quite discontinued in prescription' and, moreover, 'persons of taste or elegance seem to proscribe it even from the toilet. Perfumes, like dress, have their vicissitudes.' The olfactory tolerance of Georgian Londoners, like that of their Enlightenment Parisian counterparts, changed. Cultural historians of the senses have argued that the eighteenth century saw a substantial shift in attitudes or sensibilities towards acceptable grime and odours. Filth and smells that had been ignored, grudgingly accepted or gone unnoticed in earlier decades became increasingly unacceptable. The 'grateful' civet had been long associated with a role in disguising or complementing body odours and enhancing sexuality; this did not sit well with newer, especially bourgeois, conceptions of cleanliness and hygiene. Pungent animal-based perfumes, laden with large quantities of civet, musk and ambergris were replaced by more florally fragranced perfumes. Even where musk and civet were used, they were used in smaller quantities. Some perfumers deliberately advertised their fragrant concoctions as free from civet. As early as the 1770s, the Cheapside perfumer Richard Warren boasted his 'vegetable system of perfumery', scents without 'musk, civet, or any of those fetid drugs'. The import duty on civet too fell from eight shillings an ounce to as little as four shillings around 1800 – another sign that consumers had largely turned their noses up at civet. James Floris of Jermyn Street was the master perfumer of Regency London, and was a purveyor of the sorts of scents that people now wanted to buy: floral notes of roses, blossoms, carnations, lily of the valley, jasmine and sandalwood. Civet, so offensive to delicate minds and noses, was on the way out. This state of affairs could not come soon enough for William Mavor; to the 'credit of taste and elegance', he wrote, the 'considerable traffic in this perfume is greatly on the decline'.[9]

PART III ᷾: CROWDS

O N 25 JULY 1812 Robert Pocock, a naturalist, printer and proprietor of a circulating library in Gravesend, some 25 miles from London, was standing in a field. He thought he could hear distant low thunder. Sometime afterwards he heard a noise that sounded 'like the cough of a lion', and 'so it certainly was, for soon after several caravans passed by with wild beasts going to Strood Fair.' Mostly sailors, fishermen and oyster dredgers inhabited Strood, a village outside of Rochester on the river Medway, but for several days they were in the company of lions, tigers, a zebra, a ferocious hyena, a cassowary, an elephant and two pelicans.

In the long eighteenth century exotic animals toured the provinces as well as being a visible part of life in London. The proprietors of animals deployed a variety of tactics to draw in patrons and appeal to their sensibilities. The reception and interpretation of exotic animals – their popularity and the meanings they held – show that in Britain even the most exotic of animals like the kangaroo could be absorbed into cultural and national life. When crowds gathered to see exotic birds and beasts they interpreted and behaved towards them in ways that are surprising to the historian. The kangaroo became vaunted as a national treasure, a symbol of victory over the French. In terms of animals like the elephant, familiar to Britons long before they had ever seen one, there was also historical and cultural baggage that these animals carried. Under the theme of 'Crowds' the following seven chapters consider the ways people came to see, experience and know about exotic animals in Georgian Britain.[1]

CHAPTER 7

Ladies and Gentlemen

IN EARLY 1779 'ONE of the greatest rarities ever brought to Europe in the age of memory or man' was travelling the roads of England in a brightly painted caravan and by late January was headed towards Oxford. An advertisement placed in *Jackson's Oxford Journal* by the caravan's proprietor, Gilbert Pidcock, was intended to pique interest with the ladies and gentlemen in that city in advance and drum up a crowd. With 'the eye of a lion', 'the defence of a porcupine' and coloured 'sky blue, purple, crimson, and orange', Pidcock's wonder was appealing indeed. Literate Oxfordians versed in popular natural history would have recognized the source of Pidcock's fabulous description as Oliver Goldsmith's in *History of the Earth and Animated Nature* (1774). Or perhaps they had read Samuel Ward's *A Modern History of Natural History* (1775), which borrowed Goldsmith's description and reprinted an engraving of the 'Grand Cassowar'. The cassowary, a 6-foot flightless bird from Java, Sumatra and the Molucca Islands, was well known to the Dutch on account of their trade in the East Indies. Pidcock himself purchased his cassowary from some Dutch merchants. On arrival in Oxford the 'Grand Cassowar' was put on display at the Crown Inn in the city's Corn Market; admittance to see her was one shilling. However, Pidcock informed noblemen and gentlemen in the city of Oxford and environs that they might have the cassowary visit their house. This was a rather expensive entertainment for one's dinner guests at one guinea for 24 guests and a shilling a head thereafter. Perhaps Pidcock earned a tidy sum from his cassowary house calls, profiting from either the gentry's reluctance to visit a humble inn or their desire to impress their guests with a touch of the exotic.

Whilst in Oxford the cassowary laid a 'prodigious large egg' which attracted the interest of Oxford University's scholarly community. Pidcock had advertised that his bird had laid an enormous egg, and the incredulous but curious collegians summoned Pidcock and his cassowary for an audience. Doubting the truth of Pidcock's assertion, they had prepared a small drill with which they planned to examine the egg. The bird's green-and-white mottled egg was assiduously measured and the yolk blown to verify its authenticity in front of 30 of the university's 'nobility and gentlemen'. Pidcock made sure to advertise in *Jackson's Oxford Journal* for several weeks that he and his cassowary had received a 'warm reception' and 'much encouragement' from the scholars. The cassowary's egg attracted crowds to the Crown Inn, but interest eventually dwindled and Pidcock advertised that he would soon, regrettably, be leaving the city. His bird continued to lay eggs and for Pidcock they were, without much exaggeration, worth their weight in gold. The bird laid her green speckled eggs as she toured through Warwick, Cambridge, Reading and Abingdon, finally arriving in London. The egg that was laid in Abingdon, near Windsor, was sent to the Royal Palace at Windsor for the perusal of the King and Queen. Queen Charlotte must have been taken with her exotic egg since she ordered it to be preserved for her menagerie at Richmond, where it was displayed in her rustic thatched *ferme ornée* cottage.

The cassowary was said to, at particular times, make a 'thundering noise' which seemed to 'shake the room' like 'thunder in the distance'. The leggy bird with her resonant booming voice must have made for an impressive, if slightly fearsome, sight. Pidcock printed a small pamphlet entitled *The History, and Anatomical Description of a Cassowar from the Island of Java, in the East-Indies, the Greatest Rarity Now in Europe* (1778), which was sold as a sort of souvenir for his patrons. This type of publication was not only a money earner for Pidcock; it conferred upon him respectability as a purveyor of the wondrous bounty of nature.

Pidcock's cassowary was at that time, so he claimed, 'the only one of the species alive in England'. This was probably not mere showman's embellishment of the truth; even if it was, provincial townspeople were unlikely to be able to disprove him. His bird was of sufficient novelty at least to make a lengthy tour worthwhile and for Pidcock to mourn her loss. In the winter of 1779 the wretchedly bitter weather must have taken its toll; the cassowary died whilst on show in Durham. Pidcock was said to have considered this a huge loss on account of the 'immense sum of money' he paid for her. When Pidcock lost an ostrich he opened it up to find a stomach full of coins, glass, iron nails and small pebbles that people had thrown into the ostrich's cage. He had a post mortem performed on his cassowary too, but could not find any clear cause of death. Pidcock tried to make good on his loss and had the cassowary stuffed and later put her on display in an apartment in London along with some of her eggs. About 20 years later he had acquired another cassowary, and so in around 1800 the crowds at Pidcock's Grand Menagerie could see the living bird and one of the – by now presumably rather feather-bare – stuffed sort.[1]

When Pidcock first advertised his cassowary in 1779 he used the description from a popular natural history book because he knew that his potential audience were also readers. And if per chance they were not, then they would be impressed by the authority of the words. Presumably Pidcock, having read about the cassowary, was able to tell his patrons something of her origin and habits. In the eighteenth century books were circulated as part of lending libraries, often a service provided by booksellers. Alternatively, private readers in provincial areas would swap books with each other, eager to acquire fresh reading material. These readers were typically drawn from the gentry, clergy and affluent tradespeople, although some booksellers' records show servants and labourers as subscribers. These books circulated between readers but were not exclusively read as a private experience. Solitary silent reading is thought to have become increasingly widespread in the eighteenth century, particularly because silent reading's prerequisites – breadth of reading and prolonged exposure – were increasingly available to the middling sort and the elite. This notwithstanding, novels were still read aloud in a drawing room or salon, likewise natural histories could be read aloud and listened to. The prose and timbre of the French naturalist Georges Louis Leclerc Comte de Buffon's *Historie naturelle* (1749–88) was particularly honed to present the natural history of animals in the form of a pleasing oratory that characterized animals in dramatic and engaging ways. Read in French or in English translation Buffon's melancholy and touching description of the hauntingly lyrical dying song of the mute swan was sure to bring tears to the eyes to all assembled. Likewise the refusal of Buffon's noble elephant to reproduce in captivity resonated with the increasingly vocal rhetoric against slavery. English readers were also well acquainted with the natural histories of Oliver Goldsmith and Samuel Ward. Here too the description of the wild cassowary fleeing the encroaching habitations of civilized man, sacrificing plenitude for freedom in a barren wilderness, must have appealed to readers' – and listeners' – sensibilities and sympathies. The crowds that gathered to see Pidcock's cassowary in Oxford, many of whom would have read about the bird, might have felt sorry for her or marvelled at the ingenuity of man that had brought her from the barren wilderness to civilized Oxford.

A young male elephant billed as 'strange and wonderful' arrived at Whitefriars, London, on 3 July 1675. Like the cassowary, this elephant's novelty was advertised with only a little exaggeration. A tract printed to draw crowds to Garraway's Coffee House, where the elephant could be seen for three shillings, claimed that 'few persons amongst us, but such as travelled the Eastern World, ever saw one of them.' And it was true that, for the large part of the inhabitants of late seventeenth-century Britain, the elephant was a mythological beast. Certainly the elephant could be seen in pictorial form on painted tavern signposts, on five-guinea gold pieces, or in the pages of books such as those by Pliny or Gessner, but a live elephant had not been in the country in living memory. Yet, unlike the cassowary, the elephant had featured more heavily in cultural history and in the intellectual and imaginative consciousness of seventeenth- and early eighteenth-century Britons. This shaped the way that elephants were understood and interpreted by crowds, and the way in which they expected elephants to behave. In turn the proprietors of elephants made certain to appeal to these deep-rooted ideas to draw in a crowd.

The elephant at Garraway's Coffee House was not a well-behaved elephant; he was barely tractable and according to his keepers would 'punch either man or beast that angered him' and came within reach of his trunk. Eventually the elephant was taken to be exhibited at Bartholomew Fair, and the elephant's poor behaviour became the basis of a political satirical 'letter': *The Elephant's Speech to the Citizens and Country Men of England* (1675). The elephant's speech evoked memories of the regicide and wars of the 1640s and 1650s. The crowd was compared to the angry elephant; 'for when we are mad there is nothing that will govern us but an iron hook', but 'see at other times how tame and gentle we are'. The elephant was reminding England's citizens about the proper relationship between monarch, parliament, and citizen. The 'beast' had been willing to 'tame' its keeper in the past, but this was not a natural balance. As the elephant reminded the crowd, 'take it from me, that I never find myself better at ease than when I am obedient to my keeper'. A Royalist elephant orator was a witty and well-timed choice. The letter's writer was drawing on the classical legacy from Pliny of literate elephants competent in Greek. Given this proud heritage it is unsurprising that writers sought to couch their political critiques by putting words into the mouths of elephants.[2]

In 1683 another elephant, a young female, arrived in Britain. The tract that was printed to herald her arrival – like that of the elephant back in 1675 – strongly emphasized the relationship between kingship and elephants. It claimed that elephants could 'discern between kings and common persons, for they adore and bend upon them, pointing to their crowns'. This absolute deference to royalty was also invoked by the retelling of classical and travellers' tales of 'Oriental' kings who used elephants to administer their will. The elephant was the subject of some literary speculation and fascination in the late seventeenth century. Tyrannical kings in the East both presently and in the past were widely believed to use elephants to crush men under their feet. At the same time the elephant was not believed to be slavishly loyal to a king, since an elephant possessed a 'divine instinct of law and equity'. This elephant sense of natural jurisprudence was believed to have been evident in the refusal of King Bochus's 30 elephants to trample upon 30 men unjustly condemned to death. Crowds were thus pleased by elephant acts that showed deference to monarchs. An elephant that toured towns in the early 1700s was trained to take off his hat to politely greet the assembled company. Then he would make reverence on his knees and, after the elephant bowed, his master would ask, 'Where do you love Queen Anne?' and the elephant would point with his trunk to his heart. Such duty to the monarch went even further: on hearing her name the elephant would blow a trumpet in her honour.

Shortly before arriving in Dundee in 1706, a female elephant collapsed on the road after almost two decades walking the roads of Europe. She had been exhibited by a Dutchman, Bartel Verhgaen, since the 1680s and toured Amsterdam, Leipzig, Königsberg, Danzig and London.[3] A ditch was dug to help her get onto her feet again but this became a death-trap, as it filled with water during torrential rain and drowned the elephant. Her bloated corpse attracted the interest of locals, some of whom stole away with her forefoot. The body part had to be recovered by force. The damp English weather took its toll on another early elephant visitor too. This young male elephant arrived at Smithfield, London, in July 1720 but died a mere four months later. His short life must indeed have been miserable. His death was attributed not only to a 'want for a suitable and proportionate method of food' but also to the ignorance of his keepers. The surgeons that dissected his body thought that the elephant had been exposed to cold and moisture. A broken tusk, too, had exacerbated a fever, and 'the great quantity of ale the spectators continually gave it'. Attempts had been made to keep the elephant alive but intestinal purges carried out by a farrier probably only expedited the animal's death. In his short life the elephant had entertained his crowd by honouring King George I. He would bend his knees to the ground to 'drink his majesty's health' and paid his compliments to the company at their entrance. It was drinking to his majesty's health that had put the elephant in an early grave.

So those elephants that arrived in Britain in 1675, 1683, and 1720 came carrying the weight of cultural baggage. In addition to ideas about the association between elephants and monarchs was the classical tradition of elephants as intelligent beings. Even when the prospect of elephants actually loyal to the Crown or competent in Greek became unbelievable, they became literary anecdotes or antiquarian notes in natural histories. It is possible that some of the crowds that saw these early elephants believed the tales to be true or within the realm of possibility; indeed they liked to see elephants act like loyal subjects. But even by the 1720s this attitude had weakened. The poet and dramatist John Gay's tale *The Elephant and the Bookseller* (1726) cast a droll speaking elephant as a partner in interlocution with a drunken bookseller. The bookseller, well read in the classics, marvels at the sagacity and powers of elephant wit. The bookseller believes he has heard the elephant speak, and even 'turn a page of Greek': 'What genius I have found', he exclaims. 'Wrinkling his trunk', the elephant responded, "'Friend", quoth the elephant, "You're drunk".'

In September 1763 a young male elephant arrived from Bengal and was presented to King George III. The elephant was kept in a stable at Buckingham House Gate. Some 80-odd years since the first 'elephant letter', a new *Letter from the Elephant to the People of England* was printed. In this letter an elephant supposed that he should be appointed to a government post, and that the citizens would welcome him with such 'wonderful humility and submission'. The letter carried on, in a rather lengthy and tedious manner, to cast a satirical eye over the failings of the Georgian state. In particular, the elephant drew attention to the maltreatment of the Scots and autocracy. Here the association between monarch and elephant was used to good effect, though in this case the elephant was not such a passive subject. Queen Charlotte's elephant died a few years later in 1776 and, despite the sticky political symbolism, had continued to reside at Buckingham Gate.

The tricks that Georgian elephants performed ranged from bowing and saluting the monarch to those that pleasingly exhibited the elephant's sagacity. Blowing trumpets, waving flags, picking pockets for watches and handkerchiefs, drinking bottles of beer and carrying buckets of water were all principally intended not only to delight crowds but also to show something of the elephant's character. The most elaborate elephant performances came at the end of the Georgian era in 1830, and even here retained an air of the cultural association between monarch and elephant. For the 1830 Season two performing elephants could be seen, one at the Royal Coburg Theatre, the other at the Adelphi Theatre. The Coburg Theatre was staging a melodrama titled *Siamoraindianaboo, Princess of Siam, or the Royal Elephant*. The elephant had been trained over a period of three weeks to move in time to musical cadences and was taught to distinguish one actor from another. The latter was done so that the play's successful denouement would gratify the crowd; the elephant would 'place the crown, with true poetic justice, on the head of the lawful king'. Not to be outdone, the Adelphi Theatre's elephant would assist a 'Prince and his adherents' escape from prison by 'kneeling upon her hind legs' to form a sort of slide by which 'her friends might safely reach the ground'. The elephant as kingmaker and emergency escape slide was rooted in a long history of certain ways of seeing elephants. Indeed the pleasure to be gained from these scenes was, at least partially, predicated on a passing knowledge of those classical texts and antiquarian natural histories that had been so influential in shaping the character of the elephant.[4] By the eighteenth century, the public had long known about the elephant, if only as a foreign, almost mythological beast. The kangaroo was, however, a very new animal and quickly became seen as an important national acquisition and victory.

When HMS *Endeavour* returned from New South Wales in 1771, amongst the ship's cargo were kangaroo specimens that were dissected by the anatomist John Hunter and later painted by George Stubbs in 1773. The kangaroo was known to most Georgians only through its image and descriptions in natural histories; few had seen the scant kangaroo specimens to be found in the Leverian Museum or other collections. The public interest in the kangaroo meant that the 1789 exhibition of a taxidermy kangaroo collected by Daniel Solander attracted huge crowds; this exhibition also coincided with the publication of several travel accounts of voyages to New South Wales, all of which included illustrations of the kangaroo.[5]

The image of the kangaroo that was most well known to British readers of natural history was the one that appeared in Pennant's *General History of Quadrupeds* (1790). Living kangaroos, however, could first only be seen in the collection of Queen Charlotte at Kew from 1792, and then later in the decade at Pidcock's Menagerie on the Strand.

Georgian 'kangaroo mania' was fuelled by the Queen's kangaroos and represented best by the commissioning of HMS *Kangaroo* in 1795, replaced by a successor in 1805. At a time when there were no kangaroos on the Continent, Georgian Londoners could boast that kangaroos were exhibited in the capital at several locations. In 1800, for example, Pidcock had six in his menagerie alone, with another kangaroo held in Kendrick's Menagerie on Piccadilly. By mid-1800 Gilbert Pidcock had sold all but two of his kangaroos, a male and a female. The subsequent birth of a baby kangaroo in October 1800 drew in crowds of spectators flocking to see the widely advertised celebrity hop in and out of its mother's pouch – heralded as the 'greatest rarities ever seen'. Contemporary periodicals reported the clamour for kangaroos, and especially the baby, in London society. The 'wonderfully formed and playful kangaroos' at Pidcock's Menagerie attracted the attention of West London's fashionable beau monde and aristocracy, 'His Serene Highness the Prince of Orange, his Grace the Duke of Argyle, Lady Howe, and many others of the Nobility having in the course of the past week condescended to honour Pidcock's Grand Menagerie, Exeter-change, with their presence.' This kangaroo as a darling of London high society was quite different from the manner in which colonists in New South Wales perceived and utilized kangaroos, and the perilous voyages that brought these live animals to London.

The kangaroo was a source of meat for colonists in New South Wales and described as 'good eating'. These kangaroos were hunted with dogs or shot for their meat and pelts – or for recreation. Hunted kangaroos were also a source of the skins and preserved organs that were sent to Britain and examined by anatomists like John Hunter. Living kangaroos, especially those captured when young, could be sold or bartered and dispatched onwards to Britain. The transport ships that supplied the penal colonies of New South Wales, Norfolk Island and Van Diemen's Land with supplies and convicted transportees often returned to Britain with an exotic cargo. Anne Reed, the wife of the captain of the transport ship *Friendship*, bartered a bottle of spirits for a 'young docile kangaroo'. The kangaroo would play in the cabin and ate fruit, vegetables and bread from her hand. Later Reed noted gloomily that although she had hoped to take it safely to England, 'my poor kangaroo' had fallen down a hatch and broken its back.

The accidental death of another kangaroo is recorded in the memoirs of James Hardy Vaux (1782–1841+), convicted and transported three times to New South Wales. On a return trip to England on the transport ship *Buffalo* he noted, 'Our ship was at first so literally crowded, so as to resemble Noah's Ark.' In addition to returning convicts, the ship carried kangaroos, parrots, an emu, black swans, cockatoos and 'smaller birds without number'. But as the ship rounded Cape Horn, cold weather set in, 'destroying almost every natural production of New South Wales'. A solitary cockatoo and half a dozen swans survived the severe weather, to be presented to the Royal Menagerie at Kew.

A British presence in the new territories meant that any living kangaroo, and in fact any pelts, bones or preparations entering Europe, did so under British auspices. This control and Anglo-French conflict ensured that few other European natural historians had seen a kangaroo, dead or alive. From 1792 kangaroos grazed at

Kew Gardens, more than a decade before live kangaroos were seen in any other European collection. In Britain, however, the kangaroo had become so familiar that the animal appeared, with patriotic flair, in William Mavor's *Natural History for the Use of Schools* (1800). The kangaroo was proudly proclaimed as being first discovered by 'our British colonists' and 'unknown in other parts of the world'.

4 *A Kangaroo*, James Sowerby (1756–1822),
watercolour and gouache

Mavor's readers were told that a 'beautiful animal of this kind is in the exhibition at Exeter 'Change.'

In 1800, as British schoolchildren were introduced to the kangaroo, the Swiss physicist and chemist Marc-Auguste Pictet visited Britain. Pictet saw living kangaroos at Pidcock's Menagerie and at Kew. His detailed account of the kangaroo ('on pronounce kangarou') was published in 1802 and was one of the few Francophone descriptions of live kangaroos. Pictet's description of the kangaroos in the large breeding colony at Kew describes his astonishment and amusement at their habits and appearance. Pictet got a good look at the kangaroo in close quarters since he fed a mother and her young 'du pain' from his hand. Pictet was so taken with the kangaroos and their novel appearance that he found them to be 'a phenomenon', like a 'kind of practical joke' and claimed that 'one cannot see it without laughing'. The amusement to be had from the kangaroo was a little more brutal at Pidcock's; here Pictet observed two male kangaroos boxing in a cage and a kangaroo dancing the minuet with his trainer. Pictet found it a ridiculous and laughable scene: 'We have to say that our burst of applause encouraged him and he gratified us in putting on this spectacle.'

In 1802 Joseph Banks, during the peace afforded by the Treaty of Amiens, presented two kangaroos to the Ménagerie du Jardin des Plantes in Paris, and two years later the French Baudin expedition finally returned from 'Terre Napoléon', the southern coast of Australia, along with preserved and living plant and animal specimens. Live kangaroos were presented to Empress Joséphine, some 12 years after kangaroos had first grazed in Queen Charlotte's Menagerie at Kew. A year later and another imperial menagerie could finally boast a kangaroo as a prized acquisition: the dramatist August von Kotzebue saw kangaroos at the menagerie of Francis II at Schönbrunn near Vienna. He found the kangaroo to be 'inexpressibly comical' and thought that even a 'Cato could not forbear laughing or leaping'. As kangaroos spread through Europe into imperial collections, they were also diffusing throughout Britain; the offspring of the breeding colony at Kew were being gifted to friends and supporters of the royal family. The Dowager Marchioness of Bath acquired a male and female kangaroo for her menagerie at Longleat, as did the Duchess of York for her menagerie at Oatlands.

The kangaroos grazing on the green grass at Kew on the banks of the Thames had reproduced in captivity and clearly demonstrated their acclimatization in Britain and the ability of Britons to alter nature. Bullock's Museum on Piccadilly displayed two large kangaroos and a wallaby, and the companion catalogue to the museum considered the elegant kangaroo to be a national asset; the kangaroos at Richmond had successfully bred and rendered this 'most elegant animal a permanent acquisition to our country'. The kangaroo was a reassuring success for the Georgian state, and the touching sight of young kangaroos, kangaroos that would dance the minuet, take bread from the hand and make crowds laugh all conferred upon spectators the status of sensitive enlightened observers. Kangaroos had flourished in Britain, seemingly with ease. This appealing Enlightenment vision of endearingly docile and polite kangaroos grazing on the green banks of the Thames was, naturally, a precarious imagining. The *Monthly Review* printed a droll warning to its readers in regards to the kangaroos at Kew. It was 'not to be safe to allow them to range at large', and they were rightly surrounded by a high paling. For, although some, like the amicable kangaroos Pictet encountered, 'would allow persons to approach and touch them, especially if they give them bread', the 'others are by no means docile.'[6] Clearly the kangaroos at Kew had few qualms about biting the hands that fed them.

CHAPTER 8

Bitten, Crushed and Maimed

I N A QUIET CORNER of the graveyard at Malmesbury Abbey lies a gravestone, a memorial to the life of Hannah Twynnoy. A servant at the White Lion, a tavern in the market town of Malmesbury, the 34-year-old Twynnoy might have otherwise lived out the rest of her years in quiet obscurity. But a menagerie came to town. Twynnoy 'imprudently took pleasure in teasing' a tiger that was stabled at the tavern. Her 'dangerous diversion' continued despite the repeated 'remonstrance of its keeper'. On 23 October 1703 Twynnoy's luck ran out; the badgered and angry tiger leapt from the stable, caught hold of her gown and tore her to pieces. Thus, at the very beginning of the eighteenth century, Twynnoy had the dubious honour of being the first person to die at the hands – or rather paws – of a tiger in Britain.[1]

Britons, at least when in Britain, had little reason to fear that they would ever be slain by a fearsome beast. After all, wolves had long been absent in England and were a dwindling presence in the hinterlands of Scotland and Ireland, holding out until the 1740s and 1780s respectively. Bears, still common on the Continent, had been hunted to extinction centuries earlier. The forests, fells, marshlands and mountains of the British Isles increasingly ceased to be conceived as places of danger and foreboding. Landscape gardens, agricultural and land-drainage improvements, as well as extensive toll road building, all bolstered the impression of a managed and tamed landscape. Indeed, in the eighteenth-century large numbers of 'polite tourists', the well-to-do and the gentry, started to tour the

Lake District, the Scottish Highlands and other pleasing locales in search of a 'picturesque' scene. The landscape was tamed and Georgian Britain was free from rapacious beasts. Perhaps this made the spectacle of lions, tigers, snakes and other voracious creatures even more thrilling. Yet, as the danger from native beasts receded, a fascination with the danger of exotic animals and the gory accidents that involved them emerged, Britons overseas, working as agents of empire, still had reason to fear the beasts they encountered. In the imagination, torrid foreign climes teemed with all manner of frightful animals. The mauling of Hugh Munro by a tiger in 1792 near Calcutta particularly captured the public imagination. Sir Hector Munro, commander-in-chief of India and active in the Second Anglo-Mysore War, lost his only son to a tiger. Hugh Munro was pounced upon by a tiger and, although 'rescued from the jaws of his ravenous foe', died in spite of being attended by three surgeons. His skull was too fractured by the tiger's teeth, and his neck too torn by the tiger's claws for the surgeon's ministrations to save him. This death was, perhaps a little irreverently, commemorated by the production of a Staffordshire pottery figurine intended for chimneypieces. Munro, in red uniform, is clasped in the merciless jaws of a tiger.

Likewise, 'Tipu's Tiger', commissioned by Sultan Tipu of Mysore in 1793, is an elaborate mechanical pipe organ in the form of a man wearing the red uniform of the East India Company being mauled by a tiger. When the handle is turned, bellows inside the tiger produce the sounds of a roaring tiger and the cries of the young Englishman. Tipu, the 'Tiger', had particular reasons to commemorate the death of Munro: not least that it was Munro's father who had commandeered the forces that sacked and ravaged Mysore in the 1780s. Tipu's court at the fort of Seringapatam was painted with scenes of red-coated Englishmen being savaged by tigers. After the British siege of Seringapatam in 1799, the model tiger was shipped to London as booty, along with Mysore's gold and jewels. Once in London, 'Tipu's Tiger' went on display in 1808 at the East India Company's museum. Here the tiger organ became a much-visited attraction, a potent symbol of British ascendancy and yet, at the same time, a chilling representation of a deep-seated fear.

Animal accidents were a source of horror and a sort of morbid fascination during the long eighteenth century. These unhappy events were not, relatively speaking, uncommon – a reflection of the large number of exotic animals to be seen in Georgian Britain at that time. Bitten, crushed and maimed Britons left a historical paper trail that provides insight into their attitudes towards not only danger, guilt and responsibility but also the sorts of behaviour people engaged in around these exotic animals. Twynnoy may have been the first 'death by tiger' in Britain but hers was certainly not the first or indeed last animal accident.

Some of the earliest recorded accidents are those in the diary of John Evelyn. Monkeys, like lapdogs, were popular with well-to-do women, and Evelyn had heard of at least two instances in the 1680s in which a monkey had savaged a human baby in its crib. When at the table of Sir William Hooker, Evelyn heard Hooker's wife, Lady Lettice Coppinger, regale him with the horrid tale of a 'vile monkey'. The said monkey had 'ripped out the eyes' and 'torn the face' of a sleeping baby. The miscreant, apparently, had a taste for the flesh of an infant and was found to have 'eaten the head into the brain'. The fear of monkeys endangering children was not unfounded, or a mere hysteria. Indeed, the monkeys exhibited at the Tower Menagerie had been allowed to roam the yard until they attacked and lacerated a young boy. Perhaps Evelyn was put off his dinner by the thought of a brain-devouring monkey, though he himself was hardly faint-hearted when it came to foreign animals. In 1654 he had seen a 'huge beast' of a lion at a fair and wrote in his diary; 'I thrust my hand into his mouth, and felt his tongue, rough like a cat's.' A city gentleman with his hand in the jaws of a lion might seem an incongruous sight but, apparently, it was not.[2]

A few decades later, when the satirist and raconteur Edward Ward visited the Tower Menagerie he heard the sad tale of a female keeper who lost her arm to a lion back in 1684. The woman was said to have become the miserable object of 'her own folly, the lion's fury, and the world's pity'. Ward himself narrowly avoided receiving a lashing from a lioness's paw, upon which her keeper regaled him with a charming but jaw-dropping tale of the night he had fallen asleep inside a lion's den. He awoke to find a lion licking his face and wagging its tail in 'gratitude' for being well fed. Our proverbial Georgian 'Daniel' in his lion's den was awoken by a 'few favourite kisses' from the lion; the rasping tongue rubbed his face raw. In the 1750s another lion at the Tower Menagerie, Nero, would – so it was said – let his keeper 'play with him like a spaniel' but was otherwise a fearsome animal. This notwithstanding, it was noted that some visitors had 'been so foolhardy as to pluck a lock of his mane'. One of the lion keepers at the menagerie, a man called William, would tell visitors his own tale of sleeping in the lion's den. Instead of receiving 'a few favourite kisses', Dunco the lion lay down to sleep next to William, 'his great paw over William's breast, and laying his nose upon his head': an affectionate, if menacing, bed fellow indeed. Miss Lucy, the Tower's resident panther in the 1760s, was not quite as amenable as the lions. So a guidebook to the Tower warned its readers, 'She had recently torn the arm of a woman in a terrible manner, who attempted to be familiar with her.'[3]

When William Cowper went to the Cherry Fair in Olney, held on 29 June 1778, he saw a lion that he thought to be 'as tame as a goose'; 'I saw him embrace his keepers with his paws and lick his face.' Another spectacle proved a little too much for Cowper:

> Others saw him receive his head [the keeper's] in his [the lion's] mouth, and restore it to him again unhurt; a sight we chose not to be favoured with, but rather advised the honest man to discontinue the practice – a practice hardly reconcilable to prudence, unless he had a head to spare.

All in all, though, Cowper was impressed with the lion, it being 'much more royal in appearance than those I have seen in the Tower.'[4]

In spite of the warnings and admonishments of Cowper and other well-meaning sorts, Georgians liked to be familiar with

5 *A Lion*, James Sowerby (1756–1822), graphite

animals, perhaps because close encounters were gratifying to the senses as well as a visceral 'dicing with danger' experience. Animals that were quite ferocious were readily approached; the caged leopard at Sir Ashton Lever's Museum, housed at Leicester House in the 1770s, was 'gratified by attention and caresses'. And in return, it could be seen 'purring and rubbing itself against the bars like a cat'. A gratifying experience was a close one, and one in which the animal was responsive.

For this reason a young American visitor to London, Nathaniel Wheaton, recalled the sight of a large box of 'alligators, crocodiles, and lizards' on exhibition in a frustrated tone. They were so 'torpid and sluggish, that they scarcely deigned to move, unless provoked'.[5] Boring and dissatisfying encounters were those with unresponsive animals or a lack of interaction. Many spectators, if denied the opportunity to interact by keepers, would make their own fun. Other animal keepers were willing to facilitate hands-on encounters. In both cases, keepers needed to intervene should the worst happen. Two encounters, one from 1811 and the other from 1812, demonstrate the need for keepers to be vigilant. At Miles's Menagerie in 1811, exhibiting at Bartholomew Fair, a group of lads teased an elephant and attempted to climb on its back. The keeper, eager to show off the tameness of his elephant, had previously suffered to allow them to clamber onto it. However, this time a very unimpressed and displeased elephant spun around and squashed one of the boys against a wall in a 'very dreadful manner'. The boy, a son of a habit maker, would have surely died had the keeper not run to his rescue. These sorts of accidents were occasionally used as cautionary tales to others. Thus, in 1812 a newspaper reported a tiger mauling with the headline 'Unfortunate Accident: Caution to Others'. A man had 'imprudently ventured to touch' the paw of a tiger, which instantly pounced and after seizing his arm ventured to drag him into the den. Several men tried to liberate the man from the tiger's jaws and succeeded only by thrusting a stick into the tiger's mouth. The man was then dispatched to the surgeon, with a 'dreadfully lacerated' arm.[6]

On 19 June 1819 the boys of Dr Charles Orpen's National Institution for the Education of the Deaf and Dumb Children of the Poor in Ireland visited Polito's Menagerie, then on tour in Dublin and housed on Lower Abbey Street. The children wrote letters afterwards to recall their memories and to serve as evidence of their progress in moral and practical education. In particular the letters of two boys, William Brennan and Thomas Collins, reveal a great deal about menagerie interactions with exotic animals in the Georgian period. The boys were permitted quite freely to approach the animals and saw other visitors doing so too. After giving his 12 pennies (one shilling) for admittance, William Brennan entered the menagerie. In addition to a camel, hyena, buffalo and jackal, Brennan recalled seeing a panther. The panther was 'grinning' at a lady, and she was 'beating him with a stick'. Similarly, a man was beating a monkey with a stick and 'put the stick in his mouth'. Other children fed cakes to the elephant, an antelope, cassowary and a stork, dropping the cakes into cages. Hands strayed close to animal's mouths, beaks, claws and horns. The last animal Brennan saw was the pitiful stork, leaning crookedly in his cage. In an attempt to render the stork a more interesting diversion a man was 'beating it to get it', and then to allow the man a closer view of the wretched stork, a door porter opened the cage. Thomas Collins also teased the stork, shaking his glove in front of it, the stork responding by 'snapping his bill at it'. Later Collins took his glove and shook it at a kangaroo; he recalled in his letter how the kangaroo had jumped up to reach the glove. The letter is replete with references to touching; he shook his hand at an ostrich, saw a porcupine and 'felt his quills'. The mesmerizingly tempting sight of a sleeping lion's tail swinging down 'pendulously' from the bars of his cage was too much for Collins to resist: 'my hands were touching the tail.' The two boys did not passively observe the animals; it was a tactile experience. We should not imagine that adults necessarily warned children not to tease the animals, since as we have seen they themselves took to beating the animals to elicit a response. It is no wonder that accidents arose.[7]

John Gough grew up Kendal, in the Lake District, and would later become a natural philosopher as well as a tutor to the chemist John Dalton. Early in his childhood Gough was blinded by smallpox and his father sought ways to stimulate his son. Fortunately Gough took a liking to animals, and even in rural Westmorland travelling menageries and itinerants provided the chance to handle exotic animals. During the 1760s his father took him as a babe in arms to handle camels, bears and monkeys as they passed through Kendal. In later years, as a boy, he accompanied his father to visit a travelling menagerie and, 'an arrangement having been made with the keeper', proceeded to enter the cages. At first Gough examined the harmless animals, but this did not satisfy his curiosity. Instead, he 'begged to be permitted' to handle the rest of the collection. According to Gough's biographer, 'his entreaties prevailed' and so the young boy moved on into the cages of the carnivores. He 'ran his fingers over' all the fierce beasts, undaunted by their 'expressions of disapprobation' –

presumably growls and roars. The hyena's cage was the only one the boy was denied entry to. Though 'he was ready to make this venture' it was fortunate that 'the keeper refused to comply with his wish'.[8]

It would be easy to suppose that matters of safety and danger were neglected when it came to exotic animal exhibitions. But perhaps a Georgian would not have quite thought about it in this way. It is clear that matters of safety were given consideration, especially in relation to women and children. In this case the fairer sex and the naive juvenile warranted special protection and the reassurance that an animal was safe to approach. As such, proprietors of animals strenuously asserted that their animals were indeed safe. Such reassuring lines as 'these things are in wire cages' or 'all the animals are well secured' were commonplace in advertisements drumming up custom. Pidcock's Menagerie, the Strand's pre-eminent animal exhibition, was advertised as securing all its fearsome beasts in 'iron dens' so that 'ladies and children might see the beast collection in safety'. Moreover the beasts were 'young and pleasing', thus 'adapted to the inspections' of naturalists and the curious. We have seen that some of these 'inspections' could be quite provocative and aggravating, in which case it was fortunate indeed for curious ladies and gentlemen that they were removed from the subject of their inspections by bars of iron. Sometimes, however, fierce beasts were not confined behind bars and were instead lauded as obedient, docile and amiable. A crocodile to be seen in London in 1789 was, for example, praised for his tame and gentle nature. He would 'walk around a room silent as a lamb'. Whether or not this sweet scaly lamb took a snap at his audience's fingers and ankles is, unfortunately, unknown. Likewise boa constrictors were both excitingly dangerous and, paradoxically, pleasingly tractable. The proprietor of two boa constrictors exhibited in an apartment on Piccadilly sought to assuage the fears of his patrons. Yes, naturally an 'antipathy against these reptiles generally prevails' but the public 'may be assured that they are quite harmless'. After all, their keeper was able to play with them like a child, and as such 'even the most timid lady' might approach them without fear and danger. Whereas Mr Polito kept his

boa constrictors 'perfectly secure' in a 'commodious caravan', a keeper
at Edward Cross's travelling menagerie would – to please his crowd
– enter the cage and suffer the constrictors to 'entwine themselves
in immense folds around his neck, arms, and etc'. The constrictor
was a 'great offence to the nostrils' but otherwise charmed with its
ostensibly gentle nature. Nathaniel Wheaton found unresponsive
reptiles a bore but the constrictor 'seemed to be very tender, as he
started and gave a hiss, on laying my hand on his back'. This charming
demeanour notwithstanding, keepers such as Mr Cops, keeper of
animals at the Tower, knew that exotic animals could be dangerous.
Cops scarcely needed reminding; after all, he had been throttled and
pinned to a post by a boa – his life preserved only by the efforts of
another keeper.[9]

Mr Cops especially warned ladies that they should be cautious
of approaching his leopard too closely. The threat was not merely a
mortal one; the leopard had developed a 'particular predilection for
the destruction of umbrellas, parasols, muffs, hats, and other articles'.
Although this leopard would allow Cops to pat her, and in return
would lick his hand, she did not appreciate the familiarity of other
human company. Woe betide anyone, especially women, who came
in too close. The leopard, it was said, would seize upon these novel
items of prey and tear them apart in a blink of an eye, almost before
the astonished visitor was aware of the loss. Cops reflected that the
leopard's crimes against feminine fashion were so frequent that 'she
has made prey of at least as many of these articles as there are days
in the year.'

Many Georgian accounts of animal accidents are brief anecdotes,
like that of Mr Cops's leopard, intended to gratify readers of natural
histories with an indulgent literary flourish. Reports of accidents are
limited to short columns in newsprint or scattered broadsides. As
such, detailed accounts of accidents are scarcer, though some because
of their legal repercussions or medical novelty are well-documented.
The case of a rattlesnake bite in Regency London, the first of its
kind, is one such incident. On 17 October 1809 Mary Wombwell,
proprietor of a menagerie on Piccadilly, employed a wire-worker to

repair some animal cages. Allegedly inebriated already, 26-year-old Thomas Soper turned up at noon and set to repairing the cages in an upstairs room. Wombwell and Soper exchanged pleasantries, Soper agreeing to bring his wife for tea later that day, before the sound of customers below asking after her animals sent Wombwell downstairs to answer them. Alone in the upstairs apartment, two rattlesnakes in a cage attracted Soper's attention. Between 4 and 5 feet long, the snakes had been in the menagerie since they arrived from America the previous year. With a mind to tease the snakes, Soper took his foot-long wooden ruler and inserted it though the bars of the cage. The rattlesnakes, not prepared to play, refused to take the bait and Soper clumsily dropped the ruler into the cage. Opening the door of the cage, he hoped to retrieve it quickly, but this was a doomed endeavour. An irate and agitated rattlesnake bit him twice between his thumb and forefinger. Although Soper was drunk, the following 'incoherence of language and behavior' was later believed to be an effect of the poison. In this state he was taken to a local chemist, Mr Hanbury. The snakebite had not yet swollen, and on account of Soper's behaviour, Hanbury believed him to be simply in a state of intoxication. Soper was given a dose of jalap (a tuberous root) intended to purge the alcohol, and a tincture for his bites. Soper's condition rapidly escalated as the venom took effect.

Alarmed and increasingly incoherent, Soper was taken by carriage to St George's Hospital. Soper found himself in the hands of Everard Home, perhaps Georgian London's most eminent surgeon. For Home, a surgeon to King George III, this particular case constituted an unforeseen melancholy medical oddity. Soper's wrist and forearm had puffed-up to the extent that his shirt cuff needed unbuttoning and within the next hour or so his entire arm was swollen. Home's medical notes of his first and only rattlesnake bite patient make for unpleasant reading. Soper's death was lingering, and as such his case notes are an unhappy litany of failed medical interventions and a protracted passing. As the rattlesnake poison took hold, Soper's pulse weakened and was at points barely discernible. Over the next fortnight Home and other surgeons at St George's applied a battery of treatments intended to purge the body, reduce inflammation, numb the pain and ease Soper's fainting fits. Camphorated spirits, ammonia, laudanum and other opiates were given along, with sips of brandy. Some days into the treatment Soper's condition seemed to be improving. His stomach was able to hold down fish, veal and port. He even drank coffee for breakfast and said that he hoped even yet to still live. Cruelly the haemorrhages, ulceration and putrefaction caused by the venom worsened and Soper's state began to decline. Slipping in and out of consciousness, he 'spoke only in a whisper', an effect of his snake-induced condition and the soporific side effects of opiates. A simple prank or bored on-the-job diversion had ended in an agonizing ordeal. 'Depressed of countenance' in his last days, Soper finally died on 2 November, the day after his mortified forefinger was amputated. The subsequent inquiry into Soper's death determined that, given the circumstances, his death was the consequence of a misadventure. The verdict, 'died by the bite of a rattlesnake', placed a deodand (a fine) of one shilling on Mary Wombwell's rattlesnake. This paltry sum, scarcely even the cost of admission into her menagerie, was hardly a financial burden for the rattlesnake's proprietor.[10]

Macabre menagerie accidents were a staple of the mid-Victorian tabloid press, alongside a grisly parade of murders, suicides and other unfortunate deaths. The *Illustrated Police News* and the *Illustrated London News* in the 1860s, 1870s and 1880s specialized in sensationalist melodrama and did not disappoint their readership with their histrionic headlines: 'fearful scene in a menagerie', 'panic in a menagerie', or 'alarming accident'. Hysterical crowds, mauling lions and dashing heroes were suitably illustrated to offer a touch of verisimilitude to otherwise unbelievable tales. In the age of empire, sensuous snake charmers suffocated by their boa constrictors or dapper lion tamers torn to shreds by wild beasts made for good reading, but also tell us something about the anxieties and concerns of British society. Perhaps terrorized women and mauled gentlemen in provincial menageries were an unsettling reminder that British dominion was not total. Certainly, terrifying deaths by the claws or fangs of the animal embodiments of empire were something of a recurring theme in the Victorian imagination. Class anxieties too came to bear in these 'terrible menagerie stories': drunken or feckless working-class men who goaded and teased the animals or recklessly entered dens rendered menageries an improper place for the respectable, particularly women. Crowds ought not to press too close to cages, nor should they foolhardily stick their limbs in the way of danger.

In the Georgian period, we can see the roots of these menagerie behaviours and a keen readership interest in menagerie accidents as reported in newspapers and periodicals. The dead or disfigured included animal keepers as well as spectators. Georgians were indulging in risky business around animals long before such incidents were featured in Victorian tabloids.

'Wallace' the lion was a notorious celebrity of the late Georgian era. Born in 1812 in Edinburgh and a prize dog fighter exhibited in Wombwell's Menagerie, he featured prolifically in newsprint in the 1820s and 1830s. As Wombwell's Menagerie toured Britain so Wallace left behind a bloody trail of mangled body parts and fatalities. In two years alone he tore the hands and limbs off three people, including his keeper Jonathan Wilson; Wilson had 'imprudently and incautiously', according to the *Lancaster Gazette*, placed his hand inside Wallace's den. This folly, according to the gazette's salutary warning to its readers, was a lesson for others: 'people cannot be too cautious how they approach such animals,' after all, curiosity might be gratified without 'incurring the like danger' such as had befallen the unfortunate Wilson. George Wombwell's lion was not his only liability; a leopard too was reported as tearing at the breast of a woman and lacerating a young boy who approached the chained animal too closely. The nadir of these terrible menagerie scenes was the 'melancholy accident' that occurred in 1835 on the road outside Wirksworth in Derbyshire. A cart loaded with timber crashed into the caravan accommodating Wallace and a tiger, and in spite of vain attempts to repair the damage both animals escaped into the night. Before his capture in the morning Wallace had killed a man, in addition to several sheep and a cow in neighbouring fields. The tiger that escaped with Wallace attacked a woman with a child in her arms and an 11-year-old-boy. All died, as did the tiger, which was shot since all attempts to recover her alive were thwarted. The following day an inquest returned a verdict of 'accidental death' with a deodand of £10. The *Northampton Herald's* commentary could have cast aspersions at George Wombwell, the menagerie proprietor, for failing to repair the damaged caravan. Perhaps surprisingly, however, Wombwell emerged unscathed. Indeed, he was heaped with fulsome praise: 'too much praise cannot be given to Wombwell' since he 'expressed the utmost concern, ordered the funerals of the sufferers to take place at his expense, and promised to make good all damages arising from the melancholy event'. Wombwell settled his debts and Wallace the lion lived to fight another day; actually Wallace would die later in 1838, to be stuffed and then displayed at the Saffron Walden Museum, where he can be seen today.[11]

That the deaths of the public should be deemed accidental or hold few consequences for a proprietor or the offending animal is a notable feature of animal accidents at the end of the long eighteenth century. The deaths of the keepers who looked after exotic animals – in particular those of elephants – shed some light on the reasons juries reached verdicts of innocence or death by accident.

Tending to exotic animals was a perilous vocation, and in terrible reversals of fortune those in charge of their animals might find themselves suddenly at their mercy. As early as 1684, a woman who looked after the lions at the Tower died after a lion caught hold of her hand and 'gripped it so hard that it was forced to be cut off to prevent gangrene'. Another lion keeper walked into a den, startled a sleeping lion and lost his life. Later a different type of big cat, a leopard, escaped from a menagerie on the Strand and 'walked up Piccadilly in majestic style'. The leopard, padding through the street, 'evinced no sign of terror or alarm' though he 'produced both in every beholder'. With a fierce beast at large on Piccadilly, the leopard's keeper made strenuous efforts to catch him. In doing so the brave keeper was lacerated and his arm so severely mauled it was later amputated at St George's Hospital. At other times escaped fierce beasts could be recaptured, remarkably, without loss of life or limb.

Apart from carnivores, pachyderms with their sturdy bulky frames proved to be particularly lethal menagerie charges. A young male rhinoceros to be seen in London in the 1730s and 1740s 'bore to be touched on any part of his body'. When struck by the crowd he would anger and could only be pacified with offerings of his favourite foods – hay, sugar, rice and greens. His tongue was described as 'smooth' like a calf's and in 'running one's fingers' over the folds of his skin they felt like a 'piece of board' with the flesh between as 'soft as silk'. This rhinoceros may have been unusually accommodating and indulgent in allowing such close contact. A young keeper at Exeter Exchange Menagerie narrowly escaped being impaled by a rhinoceros after he was tossed over the animal's head. The horn passing both precariously and fortuitously between his thighs, the keeper was able to clamber to safety, the rhinoceros's horn being stuck in a wooden partition.[12] Not all keepers had such lucky escapes. John Tietjen, also a keeper at the Exeter Exchange – at that time owned by John Cross – was crushed by an elephant named Chunee in 1825 after he had attempted to chivvy it along with a broom as he cleaned its den. A tusk punctured his chest and Tietjen died within five minutes, his breastbone and ribs completely beaten in. Tietjen had only just risen from his sick bed, being incapacitated for several weeks on account of a foot injury inflicted by a leopard at the menagerie. Still troubled by his foot, Cross had told Tietjen, 'Don't go near him or perhaps you'll have your sore foot trod on.' The very day of the accident, the coroner had assembled a jury and witnesses were sworn in. Tietjen's body remained where he had fallen, and the elephant, so it was said, looked remorseful and was to be found penitently chewing on a carrot. The jury was satisfied that the death was accidental – especially since that morning the elephant had played affectionately with his keeper – and imposed a nominal deodand of one shilling. Without, it seems, premeditation, the death of John Tietjen was an accident; the death of Baptiste Bernard five years later had a more sinister air of forethought to it.

An elephant called Miss D'Jek arrived in Newcastle on 25 August 1830 to appear on the theatre stage. She had previously debuted to great acclaim at the Adelphi, and was now on tour. Thousands of curious denizens of Newcastle had crowded along the Barras Bridge to welcome her into the city. She did, however, arrive with blood on her trunk. A few days earlier, at Morpeth on the road to Newcastle, Miss D'Jek had made good on a grudge she had nursed against her keeper for several years. Whilst drunk, at some point in the late 1820s, Baptiste Bernard had stabbed her in the trunk with a pitchfork. Bernard, it was alleged, had 'ill used' her for years and as such the elephant always regarded him with 'cross looks'. Allowed to imbibe liberal quantities of beer and wine, agitated and cross elephant thoughts no doubt simmered on too. Perhaps wary of the danger that could befall a solitary keeper, Bernard was never allowed to be alone with the elephant. At Morpeth, however, an opportunity arose, and the animal gave weight to the old adage that an elephant never forgets. Miss D'Jek wrapped her trunk around Bernard's waist and squeezed him so tightly that he vomited blood and died two days later from internal injuries. On account that she had been mistreated, Miss D'Jek was spared death and given a deodand of five shillings. Her manager, Mr Yates, must have been relieved. The show could go on, with her fame untarnished by the sombre affair.[13]

This legal context is perhaps surprising; the deodand sums imposed by juries were small and owners of exotic animals were not pressurized to destroy their animals. Mauled by a lion, gouged or suffocated by an elephant, even poisoned by a rattlesnake – none of these unfortunate events culminated in conviction, or even in substantial damages. Legislation might have been expected to have regulated or arbitrated exotic animal accidents. In terms of assigning a burden of responsibility, English law certainly made owners of animals responsible for the actions of their property. Culpability was easy enough to assign where proprietors might reasonably anticipate an animal to be dangerous; moreover, menagerie proprietors could hardly plead ignorance of the true nature of their animals, *ferae naturae*. By 1800 histories of crown pleas and legal compendia had established the precedent of proprietorial responsibility, usually citing the case of Andrew Baker, a child who had been bitten by a monkey that escaped from its chain. The monkey's owner was successfully sued for damages. However, the tradition of the deodand in the Georgian period was, in part, a continuation of a long tradition of putting animals themselves on trial for murder or manslaughter. If a jury and coroner were satisfied with their innocence then gruesome incidents incurred merely nominal fines. As such, there was hardly motivation for the proprietors of exotic birds and beasts to be diligent in restraining their patrons from seeking hands-on encounters.

Of course, there need not be only a legalistic explanation for a Georgian tendency to engage in what might be considered today as reckless behaviour. We can think of the very notions of 'danger' and 'risk' as ideas shaped by cultural attitudes, as historical *things* in themselves. What was deemed appropriate behaviour around animals, 'menagerie manners', was developed over time. Spectators needed disciplining around animals and this was a protracted historical process. The 'menagerie manners' of Georgian Britons continued well into the nineteenth century. Indeed, from the 1830s the early London Zoological Gardens aspired to offer a more respectable, genteel and edifying sort of animal exhibition, open only to subscribers and guests – quite different from the rowdy and vulgar itinerant menageries or stinky animal merchants' premises of Georgian London. However, here too even polite people engaged in improper behaviour. Wandering hands, wayward canes and poking umbrellas persisted well into the nineteenth century.

# CHAPTER 9

Sweet Camel's Breath

Mr Richard Heppenstall was pleased to inform the 'nobility, gentry, and all lovers of living curiosity' that they might see his camels at the Talbot Inn on the Strand, between December 1757 and May 1758. A shrewd showman, Heppenstall rigorously advertised his camels and drew in crowds with his patter. Crowds expected a showman to tell them something about the nature or character of the animal on show, and Heppenstall obliged. Sometimes the truth might have been stretched a little or colourful antiquarian anecdotes used instead of the musings of more recent writers. Indeed, an account of the camels at the Talbot Inn that was published in the *Gentleman's Monthly Intelligencer* supposed that 'Heppenstall was very communicative, though some matters that, he says, have fallen within his observation are denied by the best writers.' This notwithstanding, Heppenstall had usefully imparted to the writer that his camels devoured five trusses of hay a week and drank very irregularly; moreover the camels shed their hair every year and were in fact shedding their hair at present. Certain preconceptions about camels as dirty or smelly creatures were the sort of public concerns that Heppenstall sought to assuage by taking out advertisements. In particular he wanted to reassure the curious women of Georgian London: 'many ladies may suppose, who have not yet seen them, that there may be something disagreeable, either to the sight or smell, on a near inspection.' Clearly, some ladies were not charmed by the idea of crowding into a tavern to gawp at a shedding or mangy camel; dirty smelly camels were not fit for the sight of the delicate and genteel.

Heppenstall went on, 'But should such supposition really exist, reasonable as it may appear, it has not the least foundation in truth.' Such vulgar misconception about the camel really ought to be dispelled in the polite and curious, after all. 'The breath of either of them is as sweet as a sow's'. It is difficult to imagine a sow's breath as sweet, let alone a camel's. But this clever comparison of Heppenstall's did something to reassure those who might otherwise have given his camels a wide berth. Particular assumptions were in place about disagreeable animal smells and appearances in Georgian culture, and Heppenstall worked to quell them. Surprisingly, it would appear that the crowds that packed into the Talbot Inn really did find the camel's breath rather pleasing. A column in *Mist's Journal* waxed lyrical about the camels: 'The ladies are especially charmed by them and express great satisfaction at the sweetness of their breath and the neatness of their apartment.'[1]

Not only were these animals themselves clean and fragrant, but their living quarters were equally tidy and pleasing. Other animals too could be described in aesthetically reassuring ways, such as the buffalo on display at the Bird Shop on the Strand. The buffalo had been acquired by Mr Brookes the animal merchant on his travels through the Ukraine, the buffalo having been brought up from Egypt by way of Turkey and the Crimea. Its odour was 'so perfectly sweet that it fills the room with a rich perfume'. And what a perfume, no doubt – the odour of manure and buffalo sweat. This buffalo, known as the 'boos potamus' or 'river cow', was also 'so extremely tractable and gentle' that even the 'most timid lady' could approach it. The lengths to which animal merchants, menagerists and showmen went to assert that their property and premises were neat and smelt good, of course, indicates by implication that many were far from salubrious.[2]

Throughout the summer of 1725 a small fleet of boats ferried Londoners to a barge moored on the Thames by Cupid's or Cuper's Bridge, a landing stage near a small pleasure garden on the Thames near Lambeth. The hulk of the barge had been fitted into a makeshift cistern, and in those murky waters swam a young and tame seal known as Toby. He had been brought to London from the

coast of Greenland by a returning whaling fleet and put on display
for sixpence. The barge was advertised as safe, especially for the
ladies, who could take full advantage of the vantage point offered
by a seating area where drinks could be ordered and taken to the
tables. The waters of the barge-cum-pool must have been filthy,
especially in the summer heat; a satirical ballad delighted in the
sordid 'evacuations' of the seal into the 'tub that was his close stool'.
The owners apparently got too greedy; they did not 'salt his tail'
and 'his flounders they forgot to bring', so Toby died that summer.
His owners had forgotten to 'give the dog his due' – 'dog-fish' being
vernacular for 'seal'. The dirty and low nature of this barge spectacle
was lampooned in the ballad that commemorated the seal, heaping
scorn on the boatloads of 'cits', or Londoners, accompanied by 'maids'
and 'whores' from Battersea and Wapping that crowded aboard the
barge. In the wake of the South Sea Bubble, a wave of speculation in
stocks and ill-fated or fictitious schemes that had swept the city into
a frenzy, the exhibition of Toby in his Thames barge was something
of a rare money-maker for the South Sea Company. Their whaling
fleet was not successful and operated at a loss. Critics no doubt
observed, as did those who penned the ballad to Toby, that the seal
provided 'rare hope of a dividend!'[3]

Like the seal's 'close stool' pool water, other animal attractions
could smell unpleasant too. When Edward Ward visited the
menagerie at the Tower of London, he thought it 'smelt as frowzily'
as a dove house or dog kennel. Worse still was the tendency of
the 'cunning' leopard to 'stare in your face and piss upon you'. The
leopard's urine 'stinks worse than a pole cat'. Most people probably
tried to avoid animal waste, not so a certain Dr William Prout.
When Prout visited a boa constrictor exhibited on the Strand, he
brought home with him some of the snake's faeces. Its scent was
'mawkish' and had a dry chalk-like texture. The constrictor poo
even left a white mark when rubbed on a hard surface. In person,
Prout observed the soft dough-like appearance of fresh excrement,
complete with a yellowish coating of crystallized urea. The keeper
explained to Prout the feeding habits of the boa constrictor and the

regularity of its bowels. His visit made the eventual 1815 publication of his *Analysis of the Excrements of the Boa Constrictor* possible.[4]

In addition to the olfactory offences of the living and their waste, it was also possible for the dead to be less than fragrant. The exhibition of a dead whale at the Lyceum in 1787 was by its very nature a temporary and odorous affair. On payment of one shilling, the 'perfectly clean and sweet' whale could be viewed. It is not clear in what respects the whale was clean or sweet; presumably the proprietor meant to reassure those who were worried that they were about to pay to view a heap of stinky rotting blubber. In fact, the proprietor did take out advertisements in newspapers to announce that the whale could not possibly be exhibited for more than 14 days at the most. Even in a frosty and chilly January, the inevitable decay could not be held at bay for long. The air in the Lyceum would have been offensive well before a fortnight had passed. Similarly, the death of a rhinoceros near Portsmouth presented a rather large and smelly problem.[5]

In 1790 Gilbert Pidcock bought a rhinoceros for £700. The young male rhinoceros had been brought over from the East Indies and presented to Henry Dundas, president of the Board of Commissioners for the Affairs of India. The rhinoceros was a present from an Indian king in Lucknow. Dundas decided to sell it rather than 'have the trouble of keeping it'. Pidcock bought it for his travelling menagerie and exhibition in London. The beast was an obedient animal that would allow strangers to pat him on the back, and he would walk around a room on the command of a keeper. Whenever the rhinoceros wanted food he would cry like a calf. He was fed on clover, ship's biscuits and greens as well as the odd tipple of sweet red wine. This he could drink in vast quantities – three or four bottles in a sitting. The rhinoceros was the subject of a painting by George Stubbs, commissioned by John Hunter. This portrait, along with the animal's stuffed skin, were by 1792 all that remained of Pidcock's wine-guzzling rhinoceros.

In October 1792 the rhinoceros, when trying to get up, dislocated his right front leg and died from the inflammation caused, around

6 *Rhinoceros*, 1790, George Stubbs (1724–1806), commissioned by
John Hunter, oil on canvas

eight or nine months later, in June 1793. He died en route in his caravan at Corsham, some 70 miles from Portsmouth, and by the time
the caravan rolled into the town the body had begun to decompose.
The stench was apparently so intolerable that 'the mayor was obliged
to order the body to be immediately buried.' He was buried on
Southsea Common, on the seafront at Portsmouth. Two weeks later,
Gilbert Pidcock returned to have the rhinoceros exhumed and preserved. But even then, 'the stench was so insufferable that it was with
the most utmost difficulty the persons employed could proceed in
their operations.' People present at the rhinoceros interment claimed
that the smell of decomposing rhinoceros was 'plainly perceptible'
from over half a mile away. Those who had otherwise wished to take
in the bracing sea air must have been sorely disappointed.[6]

Robert Jameson was downcast too, as he had hoped to see the
rhinoceros when Pidcock took up his annual pitch at London's

Bartholomew Fair later in the late summer of 1793. Jameson, future professor of natural history at Edinburgh University, was at that time a 19-year-old student at Edinburgh reading medicine, botany and natural history. At eight in the evening on 6 September 1793 Jameson arrived at the fair, only to find that the rhinoceros 'had died a short time before I came to London which loss I much regretted'. He had to make do with the sight of the rhinoceros skin, stuffed by the London taxidermist Thomas Hall. Instead, Jameson saw Pidcock's bison, 'one of the most fierce looking animals I ever saw', and a polar bear. The 'fatigued' bear was a frightful sight: 'it uttered most dreadful howls when a pole was put into it.' Jameson had been disappointed earlier too when he went to see the menagerie at the Tower of London. On 22 August he visited the menagerie along with his 14-year-old brother, Andrew, and after they paid their 1 s. 6 d. for entry they felt cheated. Andrew wrote in his journal, 'I was very much disappointed when I had seen the Royal Collection, as I had entertained hopes of seeing one of the finest menageries in London, whereas it was the most insignificant.' The best collections were those of the animal merchants and menagerists, but alas none had a rhinoceros.[7]

The importance of animal odours to Georgians reflected part of a wider cultural interest or concern about comfort and a long process of deodorization. The rejection of civet, for example, was a slow move away from pungent animal-based perfumes over the course of the eighteenth century. So too did a fear of fetid miasmas, a cause of disease and sickness, lead to a condemnation of foul smells and filth. Georgian London, like other cities of the period, did not smell good. The burning of sea coal clouded the city in smog and coated both buildings and clothes with soot. In addition to human waste, the city was polluted by the output of the scores of soap boilers, sugar refiners, iron foundries, brewers, dyers and lime burners. Even if one could not fully remove oneself from these odours, it was reassuring at least to be able to detect them, to show that one was not insensible to or ignorant of olfactory offence. A sensitivity to odour hinted at gentility or refinement and an appreciation of comfort. Historians have argued that comfort is not purely a physically 'natural' state of being, and in the eighteenth century comfort certainly came to be understood as progressive rather than natural. This means that comfort was something that could be cultivated, gained, or purchased, a perfect rationale for consumption of material goods. Likewise the cultivation of good manners, politeness and personal hygiene increased comfort. The comfort afforded by the state of one's lodgings, wallpaper and furnishings became a Georgian domestic obsession with comfort, and for the elite and middling sort gave rise to a whole vocabulary and aesthetic theory of taste and sensibility.

It was important, then, that those who wanted to make money from displaying exotic animals accommodated the taste for comfort. Heppenstall insisted on the sweetness of his camels' breath, and the neatness of their apartment was worthy of comment. Likewise Pidcock was ordered to bury his rhinoceros, for the sake of Portsmouth's residents. A perusal of the advertisements, handbills and trade cards of the eighteenth century reveals a language of comfort: animals were displayed in 'clean' and 'commodious apartments', in rooms that were 'well-ventilated' and 'fitted up' to receive visitors. In the winter, premises were also advertised as having fires lit in the interests of convenience and comfort. The hand bill of Mr Patterson, a pastry cook at 37 the Strand, printed around 1795, left the readers in little doubt that they could view his ostrich in good taste. According to Patterson, 'the largest bird ever exposed to human view' could be seen 'in an elegant and commodious room, genteelly fitted up'. The room was appropriately decorated and furnished to satisfy the demands of the discerning. It was not enough merely to see an ostrich; it should be done in stylish comfort.[8]

This comfort was, of course, not to the benefit of the animals. Premises and caravans were commodious in as much as they would admit crowds of paying customers in comfort. Georgian animal dens were unadorned pens, and usually cramped. When Robert Jameson, the student who narrowly missed seeing Pidcock's rhinoceros alive in 1793, became professor of natural history at the University of Edinburgh he came into the possession of a polar bear. The young bear had been caught on an ice floe off the West Greenland coast after its mother had been shot, and it arrived in Leith in the August of 1812. Work began on a den to accommodate him in college grounds at Edinburgh. In the interim the bear was housed in a large hogshead, or wooden cask, and had fresh water poured in daily; upon which the bear would roll around in pleasure and growl. He was fed on offal and horsemeat. The *Scots Magazine*'s natural history correspondent thought that great care should be made in constructing the polar bear's den. Naturally, it should be in a cool place and have a cistern of water for the bear to wash in. A month later the construction of the den was complete – but the aforementioned writer was not impressed: 'this den is neither so commodious nor so handsome as might have been expected from the liberality of the enlightened magistrates of the Scottish metropolis.' There was no cistern for the bear to wash in, and the den was too low and narrow 'to admit exercise'. A water pipe could have been brought in from a neighbouring well at a 'trifling expense' yet it was not. The city magistrates, on behalf of the college, wanted to save money. And on that matter 'no fund' as of yet 'provided for the support of the animal'. The 'cleanly' little animal that delighted in rubbing itself among the straw was not in sufficient accommodations nor well provided for. Jameson would not have agreed with the *Scots Magazine*'s public criticism of the bear's lodgings. He wrote to William Scoresby, the Arctic explorer who had sent him the bear, to thank him for his valuable gift. In addition to expressing his gratitude and requesting further particulars about the preferred diet of polar bears, Jameson wrote to Scoresby to say; 'I have brought the animal to the college where he is now lodged in a commodious den.' Clearly, as with the relative sweetness of a camel's breath, that which was described as 'commodious' was a matter of opinion.[9]

CHAPTER 10

Exotic Estates

O N CHRISTMAS DAY 1768, after noting with some sense of satisfaction that her husband was 'vastly pleased' with the watch she had given him, Lady Shelburne wrote in her diary about the sad death that had somewhat marred the festive cheer: 'the intense cold killed in one night our poor orangutan.' The orang-utan had lived, along with an unusually tame leopard, in the Shelburne's menagerie and orangery, both of which had been designed by Robert Adam. Lancelot 'Capability' Brown had set out the park at Bowood House in Wiltshire with an elegant curved lake, clusters of trees and a Doric temple. William Petty, the Earl of Shelburne, was a prominent Whig politician and later would serve as both home secretary and prime minister; the Shelburnes' estate was, then, an expression of aristocratic taste and cultural authority. It was also a place for the couple to host friends and guests, such as Jeremy Bentham (1748–1832), famed for his contributions to economics and philosophy; Bentham loved to unwind and write at Bowood. On 31 August 1781 he wrote to his father and told him that he was in no hurry to leave Bowood: 'I do what I please, and have what I please. I ride and read with my lord, walk with the dog, stroke the leopard.' His only obligation was to adhere to the elite habit of dressing twice a day. He returned their hospitality with a gift of a white fox that he had his brother send up from London, writing, 'Lord Shelburne is fond of collecting anything out of the way.' The white fox in the menagerie became something of a running joke among the men that visited Bowood House. They took to calling some of the ladies the 'White Foxes', probably referring to greying hair or white wig powder. It is not known what the foxy ladies of Bowood thought of their moniker.[1]

Few aristocratic menageries in the eighteenth century could boast of occupants as ostentatious as an orang-utan or a leopard; instead a collection usually consisted of assorted parrots, finches, pheasants and wildfowl, as well perhaps as deer, small monkeys and later kangaroos. Such a menagerie could be a pleasant diversion or distraction for a family or their guests; even if the menagerie was not the purpose of a visit to a house, it could be enjoyed as one of the many accoutrements that displayed the owners' taste and deep pockets. Catherine Talbot wrote to her friend Elizabeth Carter about what she had done to keep herself occupied in the 'rather autumnal' August weather of 1760. Both women were part of the social and literary circle hosted by Elizabeth Montagu. Carter rode out from London to Wimbledon to visit Wimbledon Park, one of the estates belonging to the Earl of Spencer; here she 'admired the very charming park, walked to the menagerie and all over the ground floor of the house, saw many curious and pretty birds, some very good pictures, and Mrs Spencer's closet'. The gossip about Mrs Spencer's pretty but ostentatious new decor was in fact the main purpose of the letter; even Carter's mother had apparently found the room to be decorated in a way that was 'inappropriate to the size'.

Such polite, or indeed rather impolite, guests were common to the houses of the aristocracy and gentry in the Georgian period. Even when a family was not itself present, those of genteel status might be received to admire the house and gardens. In a sense, those in the highest ranks of the aristocracy spent so little time residing in one place that, in essence, a house and its gardens were intended to be seen and appreciated by others as much as the owners. And by extension too, the feathered and furred occupants of menageries and aviaries might not have been the beloved favourites that might be imagined. For months they might go unseen by those who owned them, especially if the owners were in London or

Bath for the Season between October and June, on the Grand Tour or preferred a different estate. While the Throckmortons were away from Weston Hall, residing at their estate in Berkshire instead, William Cowper wrote to his patron and friend Mrs Throckmorton with household news. His letter told her that the clock in the hall had stopped and that 'your aviary is all in good health, I pass it every day, and often enquire at the lattice; the inhabitants of it send their duty, and wish for your return.' The aviary's wishes were granted, eventually, but not for long. A year or two later the Throckmortons removed to Berkshire for good, leaving Cowper and the birds behind.[2]

A parrot in a cage was easier to pack into a carriage or palm off on a willing friend. Elizabeth Carter wrote to another friend, Mrs Elizabeth Vesey, in July 1777 offering to take care of her parrot whilst Vesey went to her family's Irish estate: 'If you cannot find room for the parrot, it maybe sent to Bloomsbury Square with a direction to the servant who is there to send it to South Lodge.' Vesey was away from London for some 14 months, so if she had taken Carter up on her offer it would have been a long period of parrot-sitting indeed. This was a kind offer of Carter, not least because she had once been 'doomed to spend the afternoon' with a 'screaming parrot' and 'barking lapdog', along with their loud mistress and her crying baby, all crammed in a small room. A maid came in with a trumpet to shut things up. The exotic property of friends and acquaintances could also be annoying in less obnoxious ways; one could be simply bored to death on an estate with little to do but feed the birds.

Elizabeth Montagu had once mourned in a letter that she could 'not have anything of a menagerie' at her Sandleford estate near Newbury in Berkshire, since there was 'no trusting anything out of doors'. She blamed the poverty and decaying trade in the town of Newbury. Montagu had spent time 'feeding the pheasants in the menagerie' at Bulstrode, home of the Duchess of Portland. A decade later and Montagu confessed in a 1753 letter to her cousin a new hatred for birds. It seems that spending time at her friend's estate had become far less pleasurable:

I believe the menagerie at Bulstrode is exceedingly well worth seeing for the Duchess of Portland is as eager in collecting animals, as if she foresaw another deluge and was assembling every creature after its kind to preserve the species. She used to be very happy in a great variety of fowls which is a very fortunate taste for anyone who is much in the country for they have nothing to do but to throw down a handful of corn and cry 'biddy biddy' and behold their friends assemble round them in an instant while I who care for none of the winged race but your Theban swan walk alone musing on absent friends and pleasures.

Birds were the ideal companions for those starved of the company of metropolitan society and entertainment, but perhaps it was just as well for Montagu that she had not attempted to acquire a menagerie at Sandleford.[3]

Birds and animals were usually entrusted to the care of gardeners, keepers and servants, especially in the absence of their owner. The maintenance of a menagerie could be onerous, particularly if it was a large one like the Duke of Richmond's Goodwood menagerie. In the 1720s this collection included lions, ostrich, baboons, tigers, bears and a racoon, as well as an aviary. The lions and tigers ate around 70 lb of flesh a day and the other animals got through between 140 and 150 loaves of bread a week. Crowds that came to visit this unusually large collection made the job of the animal keeper Henry Foster especially difficult. Foster wrote to the Duke in his absence to inform him that, 'We are very much troubled with rude company to see the animals. Sunday last week we had about four or five hundred, good and bad.'[4]

The landscape garden at Stowe in Buckinghamshire attracted visitors throughout the eighteenth century, and a menagerie was built in 1781 for Mary Nugent, Duchess of Buckingham. Her ladyship's menagerie was a small building with a circular main room and two colonnades used as conservatories with aviaries at the end. Light streamed in through large windows and doors opening into a flower garden adorned with a fountain and statues of Adonis and Venus.

The main room was painted with a floral trellis and parrots perched among classical figures with urns. The Nugents had commissioned the artist Vincenzo Valdré to paint the frieze. Aristocratic menageries in eighteenth-century Britain could, however, be even more elaborate than that at Stowe. The menagerie at Coombe Abbey was built to resemble that at Versailles – a lodge from which animals and birds could be viewed in comfort and elegance, as part of a larger garden improved and set out by 'Capability' Brown.

From the 1750s, geometric gardens of Baroque design were largely replaced by the landscape park. Land enclosure and domestic agricultural improvement radically reshaped the countryside. By the 1780s some 4,000 landscape parks with naturalistic gardens seemingly blended into the landscape. Landscape gardeners such as 'Capability' Brown and Humphrey Repton remodelled and planted polite landscapes for their patrons, sophisticated gardens that attempted to mask or naturalize the privileges of rank and wealth. The landscape garden was a careful improvement on nature and reflected an inconspicuous sort of wealth that distinguished the virtuous and patrician from the vulgar and politically unprincipled. Exotic birds and animals – and of course foreign trees, plants, seeds and flowers – all featured in the creation of these polite English landscapes. The forced movement of inconvenient rural populations left no visible trace for sweeping eyes.

Repton thought that a picturesque landscape must hold the attention and offer a variety of perspectives to draw in the eye. In this way, a garden could be an elaborate moving wallpaper. To Repton, smoke from distant chimneys, agricultural activity in fields beyond the park, as well as grazing cattle, sheep and deer, all created interest. In particular, birds and animals created the impression of an 'animated garden'; at Woburn Abbey between 1804 and 1810 Repton used a menagerie to excellent effect. His notes describe the entrance to the menagerie from the architectural pleasure ground, walking through a classical arched door and stepping into a rustic pavilion looking out into a menagerie: 'the variety of pheasants and aquatic fowls in the menagerie creates new features in the collection of different scenery.'[5]

This sort of 'animated garden' valued both birds and animals for the impression they made, and this has some interesting implications for understanding the relationship between the Georgian elite and their menageries. The viewing of exotic waterfowl, pheasants and peacocks through a window or from a balustrade renders them as not unlike walking garden furniture, especially if one delegated the feeding of these birds to servants and was scarcely at home anyway. Viewed alongside garden follies and Chinese pagodas, these birds and animals became part of the backdrop to the social life of the estate.

Aviaries and menageries later became fashionable and affordable embellishments to the homes and gardens of the lesser gentry and aspiring middling sort. Unable to afford the master services of the likes of Repton, they could consult numerous generic scaled-down designs published in popular gardening and architecture books. Charles Loudon's *Treatise on Forming, Improving, and Managing Country Residences* (1806) considered a beautiful and elegant garden to include an aviary within 'less than a minute's walk of the drawing room'; this was obviously exotic gardening on a less aristocratic scale. Such an aviary could take the form of a glasshouse full of exotic birds in cages, with other birds free to fly around at will; wire netting could enclose an outside area of shrubs and small trees. Charles Middleton's *The Architects or Builders Miscellany* (1799) included plans and elevations for a more elaborate Chinoiserie aviary, similar to that at Kew, complete with a tearoom, stoves and accommodation for a keeper. More practical architectural assistance was offered to readers of George Todd's *Plans, Elevations, and Sections of Hot-Houses, Green-Houses, an Aquarium, Conservatories etc* (1823). Todd's plans were based on commissions executed in 1805 and 1806 in London and the Home Counties; here mesh aviaries and water pools with exotic plants were modified for town houses, suburban villas and smaller country seats. John Buornarotti-Papworth's *Hints on Ornamental Gardening* (1823) demonstrated to aspiring late Georgians how aviary and menagerie designs could be adapted to smaller estates and suburban villas: an aviary in a heated conservatory, an aviary in a flower garden to shelter one from the rain and for those with more space an aviary set on an island for quiet reading and contemplation. For Londoners with less spacious accommodation, the architecture of Georgian London – especially the new squares and developments in West London – created new spaces and prospects: verandas, balconies, porches and conservatories. All of these were used to display exotic plants and birds; indeed many thefts of birds were from the front or hallways of houses.

Menageries and aviaries on the estates of the aristocracy and gentry were probably fairly common, though rarely on the scale of Bowood House, Stowe or Coombe Abbey. Instead, collections of birds and animals were usually kept in mesh and wood pens which leave few traces. Upon the death of an owner or of the pen's occupants, these structures could be easily torn down. Some estates have suggestively named 'menagerie woods', which are probably indicative of the existence of a collection in the past. That estate menageries and aviaries were not uncommon is probably best evinced by the sheer number of London animal merchants and bird sellers. These merchants were able to dispatch animals to the country and would not have only served a metropolitan market. Likewise Pilton's Manufactory in Chelsea provided the iron fittings and occupants for both menageries and aviaries. As such, those Londoners who wanted to stock a menagerie need only go to the Strand, Holborn or Chelsea and stroll around a shop; any purchase could be sent out to their estate. This approach did have risks, however, as a Mr Saunders found in 1800. Saunders had purchased a bear, two monkeys, a leopard, a wolf and some other animals from Pidcock. They were sent to Saunders, along with the bill, in a caravan. Unfortunately, the monkeys and leopard were found dead on arrival, but Pidcock insisted on collecting payment for his stock. In court the Chief Justice, Lord Eldon, ruled in Pidcock's favour; no warranty had been provided stating that all the animals would arrive alive, so Saunders should settle his bill.

CHAPTER 11

John Bull and 'Happy Britain'

In 1806 the armies of Napoleon dominated mainland Europe and things were not looking particularly good on land for Britain and its allies. At least in terms of power on the seas, victory at Trafalgar the previous year had done much to convince Britons that the tide of the war was changing. Indeed, the dramatic capture of an ostrich and two storks destined for Napoleon's menagerie gave occasion for a little patriotic gloating. It was not until 1808, two years after, that the *Sporting Magazine* printed the full story. By then, the ostrich had settled in comfortably in its new quarters at Pidcock's Menagerie.

Two adjutant storks, large African scavenging birds, were on an American ship, the *Portsmouth* of Baltimore, somewhere off the coast of East Africa, probably headed across the Atlantic towards home. But they were 'taken', in circumstances unknown, by a French warship; in theory as a neutral power, American shipping should have been off-limits. In any event, the two storks were now travelling on a French ship, but this exotic cargo was taken in turn by an 'English cruiser'. This ship, in a strange turn of events, was later captured by the French and the intention was to send the birds to Napoleon. This particular ship had been carrying a 'fine ostrich', and Napoleon might well have happily received his trio of avian exotics were it not for the fact that the ship that was carrying them was intercepted, again, by the Royal Navy. Aboard a new English cruiser, the birds were sent to Senegal, and then on to England on the *Alexander*, commanded by a Captain Gore. One of the storks died on the way, so only one arrived alive and that was sent to the menagerie of the Duchess of York at Oatlands. The ostrich, however, was in very bad shape, and had almost been killed in the several naval battles it had endured. Musket balls had passed through the ostrich's leg and backbones; 'of these wounds it was cured' by Gilbert Pidcock. Despite its naval ordeals, the ostrich was remarkably tame, all things considered, and was described as 'so tame it will even suffer itself to be handled by the spectators'. The bird was fed on a diet of mostly cabbage and bread and was between 12 and 14 feet tall. The article that reported on the adjutant storks and ostrich claimed that it was 'one of the finest birds of its kind'. It was all the finer for having been snatched from Napoleon himself.[1]

Georgian caricaturists, on at least two occasions, portrayed Napoleon as an inmate of a menagerie. *A Full and Particular Account of the Trial of Napoleon Bonaparte* (1803) was an engraving accompanied by a text that placed 'John Bull' as judge. John Bull was the national caricature of Georgian Britain, the ruddy beef-eating everyman. Bull sentenced Napoleon to imprisonment in the care of his 'trusty and beloved friend, Mr. Pidcock; proprietor of the wild beasts over Exeter 'Change in the Strand'. There Napoleon was to be held in an iron cage, 'publicly shown to my fellow citizens'; Napoleon was later to be held in Corsica, and then strung up by his legs for life in a Mexican mine. It was an imaginative string of punishments. On the second occasion, after the actual imprisonment of Napoleon on Elba in 1814, Napoleon was once again shown as part of a menagerie – a sad-looking monkey in a harlequin coat, wearing Napoleon's characteristic bicorne hat. John Bull appears again as a menagerist, greeting spectators with a 'Ladies and Gentlemen!'

In his 1775 *Modern System of Natural History*, the Rev. Samuel Ward had counted the blessings of his fellow countrymen in 'Happy Britain'. It was a kingdom 'in the peculiar favour of heaven on thy climate; which no pernicious or rapacious animals inhabit; through which never stalks, furious with hunger, the devouring tiger; over which never hangs, threatening devastation, the voracious and unwieldy condor!' Ward's patriotic Anglican sermonizing went further in his fulsome praise of King George III. 'Happy Britain' was happier still, blessed with a benevolent 'gracious monarch' and fields that smiled with plenty; 'a gracious monarch sways thy sceptre, who never draws the sword, but in defence of freedom, and his people; who is rejoiced to diffuse blessings around him'.[2]

Britons in the 1770s certainly had reason to believe that their nation was especially blessed. In the eighteenth century Britain's colonial expansion and naval strength transformed a small European power into a mercantile empire. London, with a population of almost 900,000, was the largest city in Europe, and into the capital flowed the bounty of empire. On the ships that carried soldiers or seamen, slaves, spices and sugar also came exotic animals and birds. The dark clouds of the American Revolutionary Wars and French Revolution naturally cast a foreboding shadow over the 'smiling fields' of 'Happy Britain', but those Georgian Londoners who celebrated the end of the Napoleonic Wars in 1815 probably had much the same confidence in British, or rather English, ascendancy as Ward's generation had. In some respects, however, despite this power wielded overseas and plenitude of foreign goods, John Bull and his fellow countrymen, even those who resided in London, remained in some ways quite parochial. As much as Britons were increasingly avid readers of travel journals and guides and followers of French fashions, the world beyond Britain seemed like an inhospitable place.

The practical exigencies of war had closed off the Continent intermittently for two decades, as well as earlier in the eighteenth century. In this context the popularity of domestic tourism by the middling sorts and gentry is not surprising. The interest in national landscapes that had hitherto been considered wild became sublime and picturesque. The Peaks, the Lakes and North Wales became esteemed as peculiarly British national landscapes. Throughout the eighteenth century, Britons, and especially the English, were understood to be particularly shaped in constitution and character by their climate and landscape.

Melancholia, for instance, was thought to particularly trouble the British. The damp and overcast weather could, naturally, lower one's spirits. Moreover, the little luxuries of life such as tea and coffee and fashionable clothing took their toll on Britons' bodies, rendering them susceptible to maladies. For much of the eighteenth century corsets and bindings, on both men and women, altered the contours of the waist, chest and shoulders. Calf padding and tight dainty shoes might also serve to modify one's walking gait; we are told that Horace Walpole, man of style par excellence, entered a room 'knees bent and feet on tiptoe as if afraid of a wet floor'.

So too, by extension, would foreign travel to torrid jungles or hot deserts wreak havoc. But things were not so bad for the gloomy British, dripping in the rain; gardens, orchards and agricultural improvements had rendered them a polite, civilized and agrarian people. A resolute national character and ingenuity had allowed the people of 'Happy Britain' to prosper in a climate that, for all its faults, was at least temperate and productive. Moreover Britain, as Ward gushed, was so blessed with the favour of divine providence as to be free from all manner of rapacious and pernicious beasts. London's menageries and museums presented themselves as comprehensive collections that in some way negated the need to venture out into the four corners of the world. Pidcock advertised his menagerie as a convenience and a replacement itself for foreign travel. In 1800, when Britain was at war with France and travel was significantly curtailed, Pidcock informed Londoners that they 'never had so good an opportunity of beholding the inhabitants of the dreary deserts of Africa and Asia' since his menagerie contained 'many hundreds of rare and uncommon animals that have been procured from the four quarters of the globe'. The dangerous deserts of foreign lands need not be reluctantly explored in order to see exotic birds and beasts. All that was required was a trip to the Strand. A promotional ditty for the menagerie, also written in 1800, summed up neatly the appeal of the menagerie to Georgian Londoners:

> Were you to range the mighty globe over,
> From East to West, from North to Southern Shore;
> Were you to pass through doleful deserts,

Where neither Sun nor Moon light the hemisphere;
Or under the hot torrid zone to go
No woods, no groves, no mountains more can show to you,
Than I, in this my forest small;
Come then, one view, will give a sketch of all.

Pidcock's menagerie – his forest small – was better than anything in nature, as he could give a sketch of the entire world. A short walk down the Strand from Pidcock's was the London Museum, or Bullock's Museum. Like Pidcock, Bullock aimed to provide Londoners with a panoramic view of the world. Bullock, in his museum, presented his visitors with stuffed animals assembled together in 'natural wilds and forests', gathered as a diorama among 'luxurious plants from every clime'. The lion, elephant, giraffe and rhinoceros could be viewed as in nature, but in the comfort of the West End. Bullock claimed that his museum afforded 'a beautiful illustration of the luxuriance of a torrid clime'. This panorama was a Georgian imagining of the world, an exotic fantasy. The guide to Bullock's Museum boldly claimed that the illusion produced was 'so strong, that the visitor finds himself suddenly transported from a crowded metropolis to the depth of an Indian forest, every part of which is occupied by savage inhabitants'. In the same way that Pidcock's offered a sketch of the world without resorting to uncomfortable and dangerous travel to inhospitable deserts or jungle, so too did Bullock's offer the thrill of danger and exoticism, safe in the knowledge that the crowded metropolis was mere footsteps away.[3]

The opportunity to observe foreign birds and beasts in Britain prompted reflection on the nature of the climate; it was noted that some animals adapted to life in 'Happy Britain' more than others, like, say, the kangaroos on the Thames at Kew, which were considered to have become a permanent and glorious acquisition for Britain. The Duke of Richmond's elk were, however, a conspicuous failure at acclimatization. The Duke had intended for his male and female moose to settle in nicely at his Goodwood estate, but the patter of little elk hooves was not to happen; the male died in 1767 leaving his mate behind. The gentleman naturalist Gilbert White went to Goodwood to 'get a sight' of her in 1768 on Michaelmas Day, but unfortunately he arrived only to find she had died the morning before. White was 'greatly disappointed', especially since when he found the moose strung up in a old greenhouse the smell was already so 'unsupportable' that it precluded any attempt at a closer observation. The moose had sadly languished for some time, and the elk's keeper said that it had 'enjoyed itself best in the extreme frost of the former winter'.[4]

In the same way that the English experienced the maladies inherent to their climate and way of life, it was observed that – like the elk – other exotic animals could succumb to the melancholia and dampness. The anatomist William Stukeley's dissection of a young elephant that died at Smithfield in 1720 was published as an appendix to his greater work on the spleen. Stukeley understood the spleen to maintain a balance between solids and fluids in the body; a disordered spleen gave rise to the 'vapours' – or hysteria, melancholia and dyspepsia – otherwise known as the 'English malady'. The elephant had died from a fever exacerbated by exposure to the damp and moisture. In appending his elephant dissection to this particular work, Stukeley was suggesting that the elephant, far from its origins in a dry climate, had succumbed to the English malady.

When the exotic animals, birds and plants in Britain, especially on estates and in menageries, succeeded in acclimatizing, they enjoyed a temperate and happily fecund climate. Those that could not adapt could be sustained in hothouses with stoves, showcasing the ingenuity of John Bull. The metaphor of Polito's Menagerie at the Exeter 'Change was used by a writer to communicate to his readers the 'iron law of necessity':

Thus, as we see in Mr. Polito's Menagerie, beasts and birds brought together from all parts of the world – from the poles and the equator, under the same roof, forgetful of their original habits and their native climates, and accommodating themselves to the iron law of necessity.

The iron law of necessity was a harsh one; only a few animals would adapt, fewer still would flourish, and most merely eked out a wretched existence. Menageries – as well as museum exhibits like Bullock's – of course offered entertainment and instruction to the public, but they also represented and reinforced some deeper ideological or symbolic ideas about nature and Britishness. They demonstrated in no small degree a mastery over the animal kingdom, as well as the supremacy of John Bull's countrymen. Such control was, however, fragile and even illusory; escaped animals and menagerie accidents demonstrated that all too well. Pidcock's little forest sketch or Bullock's panorama domesticated the wild and dangerous climes of foreign lands, in the heart of the metropolis.

When the lions and tigers at the Exeter 'Change menagerie roared, the sound rumbled over the rooftops and streets near the Strand. In the thoroughfares below the menagerie, horses would startle and anxiously paw the ground. For the writer Mary Anne Lamb the lions did not represent something bloodcurdling; they became part of the rhythm of her everyday life and took on their own safer, domesticated significance. She lived with her brother Charles in their lodgings on Temple Lane, a stone's throw from the Strand. Lamb wrote to Elizabeth, the 14-year-old sister of her poet friend Mary Matilda Betham, to tell her that:

> The lions still live in Exeter 'Change. Returning home through the Strand I often hear them roar around twelve o clock at night. I never hear them without thinking of you, because you seemed so pleased with the sight of them, and said your young companions would stare when you told them you had seen a lion.

For young Elizabeth, daughter of a rural Suffolk rector and headmaster, the lions were a titbit of worldliness and London life with which to turn her provincial friends green with envy. For Mary Lamb, the roars of the lions turned her mind not to the torrid jungles or dreary deserts of foreign climes, but to matters more mundane yet closer to her heart.[5]

CHAPTER 12

Llama's Spit, a Pot of Barclay's Entire and Elephant Chops

IN THE EARLY HOURS of a morning in 1785, around one o'clock, a 20-year old bachelor, John Thomas Smith, was walking towards Temple Bar after a night out drinking. Smith was a pencil portrait draughtsman and topographic engraver, his father was a sculptor's assistant to Joseph Nollekens, the sculptor of much-sought-after portrait busts in Georgian Britain. He was born in what has been described as 'the age of portly topographies and county histories', tomes crammed with engraved plates of county seats, market towns, picturesque prospects and antiquities. As such there was usually somewhat regular work to be found and this is how Smith paid his way as a young man. Smith later became keeper of the prints at the British Museum and presided over the print room for almost 20 years. However, this was all decades away from that night when he was walking home from a night on the town. At that time he was a talented but obscure topographic engraver born into an artistic family of respectable but far from independent means. In sum, Smith was a fairly ordinary Londoner.

In the eighteenth century Wren's Temple Bar Gate, a fine arch hewn from Portland stone built during the reign of Charles II, stood on the road between Fleet Street and the Strand. Pedestrians, carts, carriages and sedan chairs passed through the gate in great volume. But Smith was met with 'a most unaccountable appearance', something far from ordinary. An elephant was being coaxed by its

7 *Study of an Elephant*, Sawrey Gilpin (1733–1807), graphite

keepers though the narrow gate. Some wharf men with tall poles accompanied the elephant, presumably to push and prod him along. Two men with ropes, one either side of the elephant, were walking down the road, charged with the difficult task of keeping the elephant as much as possible in the middle of the road. Smith saw that the elephant 'steadily trudged on' in 'strict obedience' to his keepers. The elephant was on the way to the menagerie at the Exeter 'Change, further along the Strand. Smith, in his memoirs, wrote that he would later have the honour of 'partaking in a pot of Barclay's Entire with this same elephant'. Barclay's Entire was a malty rich brown ale. Sharing a beer with an elephant proved to be an occasion worth remembering.

Sometime after seeing the elephant pass through Temple Bar Gate, Smith accompanied his friend, Sir James Winter Lake, to 'view the rare beasts' at the menagerie. The elephant's keeper assured Lake that if he gave the elephant a shilling it would nod his head and drink a pot (tankard) of porter. This the elephant did, gently taking the shilling from the palm of Lake's hand and passing it to the keeper. A beer was fetched and the elephant, 'placing his proboscis' in the tankard, drank almost all the beer. The keeper observed that only the warm beer had been left behind; 'You will hardly believe, gentlemen, that the little he has left is quite warm.' Keen to sample the dregs at the bottom of the discerning elephant's tankard, Smith and his friend took a swig to check, 'and it really was so'. A chance encounter in the street had led to another memorable one in the menagerie at the Exeter 'Change, at that time in the possession of John Clark.

Two other men wrote in their journals quite detailed accounts of exotic animals in Georgian London in the years around 1800. They recorded their impressions of stumbling across animals in the streets, animal merchants' and menageries of the city. Indeed, although John Thomas Smith's elephant arrived in London far too early to be the one that these two men saw, it is certain that James Hall and Benjamin Silliman wrote about the same elephant, or rather elephants. Gilbert Pidcock had both a male elephant and a female elephant, arriving in 1793 and 1796 respectively. In addition to the curiosity sated and amusement provided by these animal exhibitions, both men paused for thought to contemplate the suffering of the exotic inhabitants of the Georgian urban jungle.[1]

The Rev. James Hall from Walthamstow, then on the rural edges of London but now a city borough, was chaplain to the Earl of Caithness. This, naturally, gave rise to the necessity, time and means for substantial travel through the lowlands of Scotland and up into the Highlands, a journey that began in 1803. In Edinburgh, Hall found the 'collection of wild beasts' in that city to be 'but trifling': 'Why is there not a collection upon an extensive scale at Edinburgh?' Certainly, Pidcock's 'beasts and birds' in London were of advantage not only to their proprietor but also to the inhabitants of that city, being a source of 'amusement and instruction'. However, perhaps this was no small mercy for the animals that might otherwise be seen in the northern capital. After all, Hall had visited Pidcock's Menagerie and observed that he, like many exhibitors of animals, tamed them by starving. Were it not for the patrons, especially children, bringing along 'eatables', then surely the animals would 'die of want'. Hall recalled the sight of a monkey, 'poor little thing', making do with a paltry supper of one small boiled potato. This monkey was given some nuts by a well-intentioned gentleman, but threw them back at him in anger; many were but empty shells.

Hall vividly remembered the hungry elephant he had seen at Pidcock's, eagerly devouring a bushel, or small basket, of apples. The elephant had then 'sucked up' six pots of porter, one after the other, in less than a minute. The now sated elephant stuck out his trunk 'as it were, to shake hands with us', thanking Hall and the gentleman who had brought the apples along and paid for his porter. Hall, when he first saw the elephant had been disconcerted: 'I confess, I was afraid.' But, after his repast the elephant 'became extremely kind to me'. The elephant with his touchy-feely trunk charmed Hall and lingered in his mind, though he did not like it when the elephant stuck its trunk into his pocket to root out any remaining apples.[2]

'Having occupied my leisure hours, of late, in perusing Buffon, Shaw, and other writers on zoology, I have been naturally led to visit the museums and collections of animals, which are found in such perfection in London.' It was the summer of 1805 and Benjamin Silliman, a 25-year-old American, was having an enjoyable time in London. The purpose of his sojourn in the city was to buy the necessary books and apparatus for taking up his chair in geology at Yale College, as well as to attend lectures on the sciences. London, the world's largest English-speaking city, captivated American travellers. Silliman drank in the sights of the city and recorded his impressions in the form of letters to his brother, which were later turned into a published journal. In addition to the important geological matters to which he needed to attend in London, Silliman made sure to indulge to the fullest extent possible his interest in animals. He found that 'there are not many animals of importance which one may not see, at this time, in London'. On 23 July before dinner Silliman decided to visit Pidcock's Menagerie, one of those 'class of men in London who are called animal merchants'. Later he visited the Leverian Museum too, an establishment boasting a vast natural history collection. In his journal Silliman wrote that he had seen that very day, among others, a lion, tiger, zebra, elephant, orang-utan, bison, rhinoceros, elk, panther and hyena – 'most of these were living'. He heaped fulsome praise upon Pidcock's: 'this is by far the most extensive and interesting collection of living animals that I have ever seen.' He had watched the boxing kangaroos, and the marvellous elephant; it was the elephant which seemed to have left the strongest impression. Silliman threw a key into the straw, and the elephant returned it to him. Likewise when Silliman dropped his cane, the elephant picked it up. The elephant's trunk even fished out money from his waistcoat pocket; Silliman was not as easily flustered by a rummaging trunk as James Hall had been. The ladies and gentlemen in the room could not help but marvel at the elephant's dexterity. Buffon's description of the elephant's trunk, which Silliman had so recently read, had been illustrated so gratifyingly in front of his very eyes. Indeed, Silliman's journal quoted Buffon verbatim. Many of those who saw the menagerie elephant would have known, courtesy of 'Monsieur Buffon' that the elephant's trunk united touch, smell and the power of grasping, all in one member.

As he was walking back to his lodgings, Silliman regretted not seeing a camel, an animal he had not seen since he was a boy. With this regret on his mind he suddenly heard a military beat, which he thought must be a regiment of soldiers passing by in the street. But on turning the corner, 'what should I see but a camel'. His luck was in: the camel was a two-humped Bactrian, led by a man and ridden by a small monkey in a scarlet military coat. A boy mounted the camel between the humps, 'to increase the mirth'; the scarlet-coated monkey added yet more by then climbing onto the boy's head. Although Silliman had so wanted to see a camel, he could not help but notice the sorry state it was in. The camel was 'rather dispirited and poor in flesh' and reluctant to move; Silliman supposed that the rough paving stones hurt its feet. Sadder still was the proprietor's cruelty to the beast: 'he would not stir without whipping, and then uttered a piteous noise like a groan.'

The next few weeks were spent busily. Silliman went to the Greenwich Observatory to set his foot down on the line where, 'if the world will so agree longitude commences'. He saw the antiquities and geological collection at the British Museum, took a picnic on a 'delightful lawn' at Richmond and visited the Magdalen Asylum. Silliman wasn't impressed, however, by the ladies he saw on the stages of London's theatres: 'some of them are ugly enough to frighten the ghost of Hamlet.' Soon enough his mind turned again to the natural historical offerings of the city. With a special view to seeing a llama, he went to Brookes's Menagerie at the corner of Piccadilly and the Haymarket on the morning of 15 August. When Silliman entered the apartment in which the llama was standing, it was rather inconveniently looking away from him; 'wishing to have a better view I tapped him with my cane.' This did not go down well with the llama, which promptly 'flew into a violent rage', turned around and 'ejected from his nose a greenish fluid into my face'. Silliman retreated but the llama could not be easily placated; every attempt to 'conciliate the animal's favour only produced a fresh shower'.[3]

The piteous groans of the whipped camel moved Silliman, but not to the extent that he had second thoughts about 'tapping' the llama with a cane, impudence for which he received his just deserts. The poorly fed animals, heavily dependent on handouts, no doubt prompted Hall and his friend to bring along apples, and yet the tankards of Barclay's Entire did more to line the pockets of Pidcock than nourish the elephant. Likewise, we have seen how keepers and patrons alike routinely beat menagerie occupants. The widespread nature of this neglect or violence reflected a general indifference in Georgian culture to the fate of animals. This is not to say that sensibilities were not softening, but rather that animals occupied a very different place in the cultural imagination from the one they held in the later nineteenth century. Few tears were shed for animals; those that fell were literary, satirical or reserved for lapdogs and parrots.

The shocking death of Chunee the elephant, a celebrity animal for late-Georgian Londoners, did, however, cast into relief the plight of exotic beasts in the city. Chunee had been in the Exeter 'Change's menagerie since 1814, and in good times was tractable and did tricks for his audience. His keepers could sleep in the den next to him. At the worst of times he was dangerous and violent, spurred on by his 'must', a seasonal desire to mate. As we have seen, Chunee had killed his keeper in the past, John Tietjen, and by 1826 even heavy doses of Epsom salts were no longer sufficient to subdue him. Moreover, the little elephant that had been coaxed up two flights of stairs and put in his den was now, some 16 years later, a big angry elephant. The floor of Chunee's den had been reinforced with bricks and wood to prevent the collapse of the elephant through the floorboards and into the shops and arcade beneath the menagerie. However, the flooring elsewhere in the menagerie was not elephant proofed; any escape would almost certainly lead to structural damage. The worst-case scenario, one that was by no means far-fetched, was an escaped bad-tempered elephant crashing through the floor and stampeding down the Strand. On 27 February 1826 an enraged Chunee tore apart his den, seriously damaging the iron and wooden fittings; his cage was now barely holding him captive. Men had to be found and supplied with pikes to keep him from charging and bringing the heavy door off its hinges. His proprietor, Edward Cross, needed to take action and procured 4 lb of arsenic from Mr Grifford, a chemist in the Strand, planning to poison Chunee. The arsenic was mixed with oats and sugar but Chunee would not touch it. Earlier Cross and his men had also tried to trick him with sedative-laced oranges, his favourite fruit, but Chunee had stomped on them. Mrs Cross had thought of feeding him buns with calomel (mercury), using a boy to feed him lest the elephant's suspicions be roused. Chunee ate the buns, leaving only the bun with calomel inside. He stamped on that too. Mrs Cross was apparently fond of the elephant, but at this stage she knew that they had run out of options.

As a last resort the doors to the Exeter 'Change were closed as Cross prepared for the worst. Soldiers and keepers armed with muskets shot at Chunee in a grim and gory standoff. The sight was pitiful as the elephant's 'shrill cries of intense agony' filled the air. But 'this was not the moment for the indulgence of kindly emotions'; Chunee was to be dispatched, even if it meant resorting to brutal means. Cartmell, one of Chunee's keepers, kept himself as hidden as possible so that when he commanded Chunee to kneel, the elephant would obey his voice. When kneeling, the elephant's head and throat were an easier target; even so it took over 150 musket balls and an hour to kill him.

Cross, consummate showman that he was, turned a tragedy into a money-earner. The public could pay admittance to view the grisly scene. In time, Cross sent for the butchers and they were accompanied by eager anatomists. Using a pulley attached to a crossbeam, Chunee's body was lifted and then flayed. Some of London's most eminent anatomists and surgeons were in attendance. First his trunk was cut off; the trunk that had picked coins out of the palms of spectators and then sucked up beer. Then his body was cut open to extract the internal organs for dissection. The anatomists present, apparently, had never viewed 'a more beautiful display'. His skeleton was sold for £100, although it was cleaned and exhibited first at the menagerie and then in several museums before going to the Royal College of Surgeons; his musket ball marked skeleton remained on display until 1941, when it was destroyed by an incendiary bomb during the Blitz. Chunee's keeper had once climbed a ladder to rub oil into his dry cracked skin; in death that skin was sold to a tanner for £50. The 'stench' of the local neighbourhood was said to have been 'very offensive'. Around 4 tons of flesh was sold to butchers, who loaded the elephant meat onto carts. Chunee had become cat meat. Rumours in newspapers suggested that two large steaks had been cut off Chunee and eaten at the menagerie. Some said that 'elephant steaks' had been served by Mr Cross to a 'party of friends'. Others claimed that 'rump steak' had been eaten by the anatomists and been described as 'a fine relish'; a satirical verse imagined 'rioters all feasting on elephant chops'. The actual fate of Chunee's flesh remains uncertain, but perhaps a little of the 4 tons of elephant flesh sold that day did end up on the dinner plates of a few Londoners after all. The Chunee incident gave considerable traction to the idea that cramped menagerie conditions were inadequate, even cruel. Those who advocated for a national menagerie or zoological garden, a respectable collection out of the hands of menagerists and animal merchants, had in the example of Chunee a compelling case in point. As it would turn out, the demise of Chunee was rapidly followed by that of Cross's Menagerie itself at the Exeter 'Change.[4]

The anatomists and surgeons who dissected Chunee were not the first to have worked on an elephant in Georgian Britain. In fact, the elephant had first been anatomized in the 1680s. Exhibited exotic animals played an important role in creating natural historical knowledge. When living, foreign beasts and birds were traded as luxury goods or exhibited in menageries and animal shows. In death, however, they lived an afterlife as anatomical exhibits and drawings in museums and libraries. Following the lives and afterlives of elephants in the eighteenth century sheds light on the production of knowledge about exotic animals by anatomists as these exotic beasts went under the knife.

CHAPTER 13

Under the Knife

I F YOU HAD BEEN walking along the Thames at Whitefriars on 3 August 1675, you might have been fortunate enough to witness the arrival of a living elephant. The advent of this 'strange and wonderful' spectacle in the city drew Londoners and scores of 'admiring countrymen' who were 'always greedy of novelties'. More familiar with seeing the elephant on 'signpost in wretched painting', it was without exaggeration that the young elephant's proprietor Mr Wilkins could boast, 'few persons among us, but such have travelled the Eastern World, ever saw one of them.' This elephant must have been an astonishing sight for late seventeenth-century Londoners, not least because his keepers reported that he was liable to use his trunk to 'punch either man or beast that angered him, or came within his reach'. From London, the elephant began a six-year journey through the towns and country roads of the kingdom, reaching Dublin in 1681.[1]

In the early hours of the morning of 17 July 1681, the wooden booth in which the elephant was stabled caught alight. In a frightened state he attempted vainly to escape, breaking his right tusk in half. Burnt alive, the elephant left behind grim charred remains. At dawn a large crowd gathered and, so we are told, 'endeavoured to procure some part of the elephant, few of them having seen him by reason of the great rates upon the sight of him'. But a chagrined Mr Wilkins was not going to let Dublin's trinket hunters carry away what remained of his livelihood. He hired armed men to guard the carcass and retrieve the stolen remains by force. By nightfall the next day, the elephant was creating an 'intolerable stink' that could be detected in the nearby Customs House and Council Chambers. Fearful of incurring the wrath of the Lord Mayor and Lord Lieutenant, Mr Wilkins decided to send for butchers to dispatch the elephant. An Irish physician, Allen Moulin, 'delirious to inform' himself of the structure of the elephant, joined the butchers. That night the first dissection of an elephant in the kingdom took place, with the boiling of bones and the cutting of charred or parboiled organs carried out by candlelight. So ended in tragedy the Whitefriars elephant's journey through 'our uncouth country'.[2]

The elephant mentioned earlier in Chapter 7, which died on the road to Dundee, was later dissected by Patrick Blair, a Dundee surgeon and apothecary, who presented his findings to the Royal Society and Hans Sloane. Sloane would, later, have the rare opportunity to dissect an elephant for himself.

That elephant arrived at West Smithfield on 2 July 1720, but it did not live long in London. A short few months later *Mist's Journal*, the same publication that had announced the elephant's arrival, reported the animal's death. His sad end has already been mentioned as the result of a combination of the damp and cold, a fever, and the ineffectual ministrations of his keeper and a farrier.

When these three elephants were alive, crowds had gathered around them on payment of an entrance fee. In death, the elephants began an afterlife with a different and more limited crowd, the circle of anatomists and naturalists. The dissection of body parts and their preservation created a whole range of new objects that were circulated to different collections and 'written up' as anatomical accounts. The elephants had wandered the muddy roads of the kingdom in life; in death they moved through anatomists' collections and cabinets as bones, preserved organs and skins. These were used, along with the observations of elephants made in life, to challenge existing received knowledge about elephants and to create new ways of knowing the elephant. Moulin, for example, who had dissected the elephant that died in the Dublin stable fire, kept the material he dissected to examine and use to write his anatomical account of the elephant. He reassured the Royal Society that he had done everything possible to recover and preserve as much as possible from the burnt elephant carcass. Likewise, Sloane and Dr Douglas removed organs for further examination; Sloane stretched out parts of the elephant's brain onto a sheet of paper, whilst Douglas took home those organs 'pertaining to generation in a female'. When Patrick Blair wrote his anatomical account of the elephant, he recorded not only his own process of preserving and mounting an elephant skin and skeleton; he also noted the way in which Moulin had done so 25 years earlier. Blair's Dundee elephant, drowned in a ditch, had deteriorated from exposure to the elements, so his priority was to salt the skin and preserve the skeleton. Moulin's elephant had been burnt in a fire and the skin was incinerated, thus making Blair's preservation of the skin all the more important. Blair published his 'Osteographica Elephantina' with copperplates showing the skin, mounted skeleton and bones of the elephant. The stuffed skin stood in the hall of the Royal Society, a building on Crane Court off the Strand. Blair considered it a job well done; the skin was, according to him, 'done to a good purpose', 'lively' and a 'most curious ornament'. The elephant's skeleton was mounted and kept in Dundee in a repository. Blair thought that Moulin's elephant mount was a poor one; the skull was placed incorrectly, and clumsy ironwork obscured the skeleton. Blair therefore decided to put an iron framework inside the bones; running iron rods through the spinal cord, and wiring the skull and feet so that none of this was visible to the beholder. He even used beaten and wetted leather to create replacement cartilage and ribs. His stroke of genius was to turn the elephant into an articulated model by pinning the jaw so that it could open and allow a full investigation. Likewise the skull was cut so that it could be opened to reveal the inner structure. In life the elephant had been on show in a wooden booth, now it was on display in the Dundee Repository of Rarities, near to where she had died. That the Royal Society in London should get a stuffed skin and an obscure repository in Dundee kept the skeleton is perhaps somewhat odd. After all, this repository was not eminent enough to leave much of a trace; indeed, it is known only through this elephant gracing its halls. Blair must have been closely involved with the repository and, as the gentleman who had dissected and prepared the elephant, he was able to dispose of it as he saw fit. The elephant bones at the Repository of Rarities were said to have been eventually ground down to make a top-dressing for the fields of a farm in Strathmore.

Access to specimens in collections and cabinets was important because they made possible investigations that would allow publication and furtherance of one's status in the Royal Society. The dissected and reassembled remains of elephants became a resource through which to understand the elephant and make sense of previous classical accounts. Elephants were not new in the seventeenth and eighteenth century, in the sense that knowledge about them had percolated around for centuries. But they were new in that they had never been observed so close at hand, nor gone under the knife. Anatomists at first tried to ground their findings in classical authority, couching their ideas in those who had gone before. The elephant anatomical writings of Moulin, Blair and Stukeley show the process of making new animals out of a mythological and classical beast.[3]

Pliny's writing about elephants heavily influenced Moulin's interpretation of the few burnt remains of elephant skin in his possession. Moulin had probably read Pliny's *Naturalis Historia*, which had been translated and published in 1601. Moulin saw the grey wrinkled skin of the elephant in the same way as Pliny, as a means of crushing annoying flies by contracting the wrinkles together, squishing the insects between skin folds. Blair, however, disputed this, since he had observed elephant skin through a microscope and taken careful measurements of the trunk and tail. He reasoned that an elephant's trunk and tail made a far better defence against flies; elephants swatted flies rather than squashing them. The old belief that elephants were afraid of mice was one that Moulin sought to explain through his own observations and those of the elephant's keepers. Since the elephant lacked an epiglottis, maybe elephants worried that a mouse might crawl into their trunk and suffocate them. This seemed reasonable since the keeper had seen his elephant sleep with his trunk close to the ground, so that only air might go up it.

Stukeley used antiquarian sources to underscore anatomical observations too in the 1720s. Pliny's tale of a learned elephant who could read and write in Greek was well known in this period; Stukeley invoked it to describe the elephant brain and call attention to a sense of wonder at its structure, 'so fine and perfect, that we need not wonder this creature, according to history, should be the wisest of all beasts, and even imbued with human passions'.

The menagerie acts of the elephants were popular because they gave life to the great qualities of the elephant, proving their sagacity, and audiences who knew – if only in part – of some of the classical and antiquarian elephant tales, were sympathetic to this kind of way of thinking about elephants. Similarly, early anatomists, even when they doubted classical authority, sought to ground their new observations by referring to those of the past as antiquarian curiosities. This trend persisted well into the eighteenth century; Tennant's *Natural History of the Elephant* (1777) represented the elephant as a wonderful beast with prodigious talents, drawing heavily on antiquity and travellers' tales.

Some far-fetched elephant tales were difficult to disprove and persisted. In 1674 the pamphlet that was printed to accompany and inform crowds of the 'strange and wonderful elephant' enlightened readers on the delicate matter of elephant sex. Apparently the elephant was a fairly chaste animal, taking 'venereal compliments' infrequently and in a private place. A she-elephant would recline on her back, on a bed of herbs, whilst the male covered her. This was an idea that circulated widely. Patrick Blair thought the idea that elephants embraced in the missionary position was an appalling one, an inversion of an ordained natural order. In any case, an elephant was far too bulky to hunker down upside-down on a bed of greenery. Yet this image of elephant sex hung around and was one that 40 years later William Stukeley and Dr Douglas subscribed to. Classical authors generally believed that elephants waded into pools of water when they wanted to give birth; a male would stand by the pool to guard the mother from their mortal enemy, the dragon. Dragons and unicorns had appeared in natural history texts until the seventeenth century, though by then their inclusion was more one motivated by antiquarianism. Clearly the early eighteenth-century anatomists felt as though they needed to select prudently and with a pinch of salt the classical sources upon which they chose to base some of their anatomical observations. As greater numbers of live and dead exotic animals arrived in Britain, and especially in London, anatomists or physicians in the last half of the eighteenth century had more material on which to base their observations.

Queen Charlotte had two elephants in her menagerie in the 1760s and 1770s, alongside her zebra. When the first died he was sent to the anatomist William Hunter, physician-extraordinary to the Queen. His brother, John Hunter, became surgeon to George III in 1776 and was able to dissect the Queen's younger elephant when it eventually died. John Hunter was one of London's most eminent surgeons and anatomists, with his own school of anatomy and collection. The elephant went on display at the Leverian Museum in Leicester Square, opposite Hunter's anatomy school and London residence. Many of the exotic animals brought to London, sooner or later, went under John Hunter's knife; he was well known to animal dealers and keepers in London, and would often have first refusal on an animal. This was a privilege he enjoyed, for example, at the Tower of London Menagerie as the King's surgeon. Hunter would even borrow money from friends to pay for animal corpses to dissect. He also bought live animals, but then turned them over to others for public display, content to dissect them on their death, and no doubt avoid the costs of their maintenance. His country house at Earl's Court had a menagerie with leopards, lions, buffalo, wolves and mountain goats. Mr Brookes, of Brookes's Menagerie in the Tottenham Court Road, presented Hunter with a puppy that had been born from a wolf father and a Pomeranian mother. That such a fierce animal would breed with a tiny lapdog was mind-boggling and unusual. Hunter's puppy grew into a dog with quite unusual tendencies, more wolf than dog. This wild character led to the dog being stoned in the street for a rabid dog. Hunter was on good terms with London's animal merchants, and Mr Gough, of Gough's Menagerie, recalled the ease with which Hunter would allow him into his home; even when his house was full of patients and carriages waited by the door, Hunter would admit him, saying: 'You have no time to spare, as you live by it. Most of these can wait, as they have little to do when they go home.' Hunter's patients were the wealthy and titled, but he made sure to make time for busy animal merchants with business to attend to. Hunter also benefited from a good relationship with Gilbert Pidcock: Hunter commissioned George Stubbs to paint Pidcock's young rhinoceros in 1790, the poor rhinoceros that would later dislocate his leg and be temporarily buried on Southsea Common, Portsmouth in June 1793. Hunter's menagerie, along with his vast collection formed through access to animal cadavers, helped to consolidate his position as Georgian London's foremost authority on matters of animal anatomy.[4]

Joshua Brookes, anatomist, born into a family of animal merchants, also kept a menagerie at his house and anatomy school on Blenheim Street. He amassed a large collection of human and animal anatomy, but his anatomy school suffered when surgical education became more tightly regulated. Brookes was forced to sell off his collection, and it was eventually sold at auction in 1828 and 1830. Brookes was the anatomist who had dissected the famed parrot singer Miss Poll and had kept her in his collection. His bird and animal merchant siblings furnished his collection with specimens too; the catalogue to the sale of Brookes's collection lists some of the animals he received from his brothers. Paul sent him a flamingo, tapir and crane. Joshua noted the donor as 'Paul Brookes, Esq'. Joshua had also attended to the dissection of Chunee the elephant at the Exeter 'Change, along with some students; and if the rumours were true he partook of some elephant steak. His menagerie was a rock grotto, with a fountain, upon which birds and animals were chained. A fire at a theatre on Great Marlborough Street spread to other buildings nearby on 14 January 1792; the heat from this inferno warped and blistered the door and window frames of Brookes's house. A crowd could see the menagerie in chaos through the wide iron gates. The terrified animals and birds were pulling at their chains and panting: 'the mob assembled, and fancying that the poor animals were roasting alive, kept up an alarming yell and threatened to pull the house about his ears.' The mob saved the animals, and didn't tear the house or menagerie down; Brookes kept the grotto for many years, almost four decades, until it was dismantled for sale in 1828–30.[5]

As exotic birds and animals were observed and dissected, London and provincial anatomists or physicians were able to bolster their reputation as authorities, especially if they were able to acquire and collect specimens and report their findings to the Royal Society. The city's network of animal merchants and menagerists was vital to the practice of anatomy, so for men like Hunter and Brookes it paid to be on good terms with those who dealt in animals or owned them.

PART IV ❦ HUMOUR

EXOTIC ANIMALS AND BIRDS permeated Georgian culture: in humour, satire and erotica. These animals in eighteenth-century London, and in Britain more widely, occupied not only the places in which people lived and worked, but also their minds. The foreign birds and beasts were used to represent or satirize those who owned them, or the presence of these animals could provoke cultural concerns about luxury or inappropriate sentiments. The following three chapters show the ways in which foreign animals and birds could be thought and laughed about as they became entrenched in Georgian ways of thinking. Some of this humour was very masculine as erotic eels and the idea of parrot lovers titillated gentlemen; women told other jokes about women, as ladies of a certain sort dreaded being thought of as 'a parrot'.

CHAPTER 14

Electric Desire

F ROM THE MUDDY RIVERS of Dutch Guiana to a sensational living electrical spectacle on the Strand, the journey of five electric eels to London in 1776 is a remarkable tale. The story of the electric eel in Georgian London intersects with the histories of electricity and erotica and is a story that began on the trade routes of sugar and rum. London's well-to-do, including anatomists, electricians – experimenters with electricity – and connoisseurs of erotica, crammed into darkened rooms to see the 'eel's electric flame' or experience the 'electrical spark' course through their bodies. The presence of these eels inspired a great deal of erotic humour and satire; not least when they finally died, exhausted after a long season of experimentation and shows. The eels had been imagined as erotic spectacles, and their loss was mourned:

> Ah dear Gymnotus! Pride of all the land,
> Joy of my heart, and partner of bliss;
> I've seen thee oft magnificently stand,
> And shared with thee the rapture of a kiss.
> That eel which made the very dullest rise,
> Is robbed of rigour and electric flame.[1]

Before the eels had been shipped to London and robbed of their rigour and electric flame, they had lived in muddy rivers in the Dutch colony of Guiana. This colony was part of the Atlantic trade in slaves, sugar and rum, and this trade route was the means by which the eels came, first, to the American colonies. The British government wanted to compel Anglo-American merchants to buy only British rum and sugar, thus paying the attendant excise and duties and sustaining a mercantilist trade balance. Of course, some merchants sought to evade these restrictions and pursue an illicit and lucrative trade in Guiana rum, shipping it into New England and then on to London and the European markets. The ship of Captain George Baker was one such vessel that plied this trade route, and Baker would play an instrumental role in bringing electric eels to London. He managed to bring some eels from Guiana into the American colonies – quite a feat, as it was difficult to capture and keep these eels alive. Electric eels had not been shipped alive into London before, but eels of the more common sort had, and in huge quantities. Eels could be fished from the Thames, but demand outstripped supply; thus Dutch ships brought live eels into London to sell at Billingsgate.

Once he had caught his electric eels, Baker first took them to Charleston and Philadelphia during 1773 and 1774 and made a handsome sum of money showing them to a colonial audience; but Baker had his sights set on a more lucrative market and decided to dispatch some to London. His first attempt was met with failure, as none of the eels arrived in London alive in the winter of 1774. Although Baker was no doubt bitterly disappointed, the anatomist John Hunter was delighted to receive fresh specimens to dissect.

Finally, Baker was successful and brought living electric eels to London on his second attempt. He claimed that he had caught these eels himself and that it was with much strenuous exertion that they had arrived alive. The physician and chemist Edward Bancroft described the difficult work of capturing electric eels and keeping them alive. When he was in Guinea, Bancroft had seen the preparation made for capturing and handling electric eels prior to shipping. Eels had to be young to be caught for raising in large troughs of freshwater, made especially for the purpose. A trough required scouring clean every day, as well as a change of water, since the eel's skin 'secreted a slimy substance' that fouled the water. Whilst the trough was being scrubbed clean of slime, the eel or eels would lie motionless as they were exposed to the air; yet even in this vulnerable position they were still capable of giving a shock. The eels were fed on a diet of small fish and cockroaches, though earthworms were an acceptable substitute. A lack of fresh water onboard a ship meant that a trough could not be cleaned so easily of slime, or refilled. The rolling motion of the ship on the high seas inflicted bruises on the soft-bodied eels. Given the perilous journey the eels had to make on their passage to London, it is not surprising that several earlier attempts had failed. Baker succeeded against the odds in August 1776. The chances of live 'Gymnotus Electricus' making it across the Atlantic were significantly lowered by the disruption caused by the American Revolutionary War from 1775. The Royal Society had been waiting impatiently for the eels to arrive for some time.

Alexander Gardener in Charleston had sent the Royal Society notes on the electric eel, informing them that it was impossible to accurately inspect a living specimen on account of its 'amazing power of giving so sudden and violent a shock' to anybody who touched it. Hugh Williamson's report from Philadelphia gave a less bleak opinion; indeed, his experiments gave an insight into the great research potential the eels could provide. Williamson's experiments included throwing catfish into the water for the eel to stun; attempting to block the shock with different materials like cork, brass and silk; and dipping his hand into the water to provoke the eel into shocking him. The most impressive of Williamson's experiments was when he gathered a company of eight to ten people and had them stand in a circle, whilst holding hands. With a person grasping the eel's head, and another the tail, an electric shock was administered to every person in the circle.[2]

In the eighteenth century the electric eel was compared to a Leyden jar. This was a device that contained an electrode and electrostatic generator and had been developed by the Dutch in the 1740s. The Leyden jar was used in a range of dramatic public demonstrations of electricity, and electrical showmen toured Europe to give electric shows in darkened rooms for paying audiences. Such a room was lit only by the electrical charge. The most famed of this sort of electrical soirée was when a lady would stand upon a stool as a charge was generated; a gentleman would try in vain to kiss her but was repulsed by the electrical charge carried through the lady's body; the charge would go through the stool and her whalebone corset, reaching her lips. As the lady quivered with electricity, flashes of light were emitted from her corset. This trick was known as the 'electrifying Venus'. The touring electrician Benjamin Martin could produce spectacular 'purple fire' as well as a 'beautiful apparition of stars' using his glass globe that contained electrified thread.

Other uses for electricity in the Georgian period were less frivolous. Electricity was increasingly used to treat maladies such as muscle cramps, gout, kidney stones and a cessation of menses. Treatment involved sitting on a chair or bed and clutching a wire connected to an electric machine, some form of modified Leyden jar. Edward Bancroft claimed that physicians in the American colonies, Surinam and Guiana had experimented with the use of electric eels as a cure for paralysis. In London between the 1770s and 1790s electricity began to be first used to revive the apparently dead or drowned. Humane societies charged with aiding accident victims sometimes took up these devices, and although the use of electrical resuscitation remained limited in Georgian London, some lives were saved. Electricity could also be used to kick-start a waning libido; some Georgian physicians believed that electricity could increase virility. These sorts of treatments and ideas infused electricity with an air of sexual tension, life-giving mystery and spectacular showmanship.

Once they finally arrived in London, George Baker and his five electric eels first set up at the Haymarket, but later moved to an apartment along Piccadilly opposite St James's Street. Baker's show was located to take advantage of the deep-pocketed residents of the beau monde residing in St James's and Mayfair. The debut show of the electric eels was a high-profile society event attended by 70 eminent figures, including the Duke of Devonshire, the Marquis of Rockingham, Lord Cavendish and the president of the Royal Society, along with around 30 of the society's members. George III and Queen Charlotte also patronized Baker's electrical show. The experiments that Baker carried out were broadly similar to those that Hugh Williamson had conducted in Philadelphia. The Italian electrician Tiberius Cavallo was resident in London from the 1770s and he wrote about his experience at Baker's exhibition.

One could receive a mild shock by putting a finger or hand into the water in which the eel swam. But the 'best way' to get a stronger shock was to grasp the eel firmly by its head and tail. This Cavallo did, and felt a powerful shock in his arms and chest. Other participants in such demonstrations received numb limbs, were brought to tears, or lay prostrate on the floor to recover. Sometimes the numbness lasted for several hours. The experiments were so popular that Baker had to inform the public that, regrettably, the performances would be reduced. He was 'obliged to draw the spark or vivid flashes from [the] fish but three times a week that is Mondays, Wednesdays, and Fridays, on account of the danger of their being exhausted by the too repeated experiments of everyday'. The full range of electrical experiments could be seen for the large sum of five shillings. The daily wage of an unskilled labourer in London was about one shilling, meaning that the extravagant eel show was too expensive for many Londoners. The cheaper rate was two shillings and six pence, and this entitled a person to feel an electric eel anytime between ten in the morning and four in the evening. With such an intense schedule of shows, Baker's eels fairly quickly succumbed to sickness and exhaustion. As a result, Baker was forced to scale back his show and entirely discard the dramatic but draining experiments. Grabbing the head and tail of the unfortunate eels in order to feel a strong shock was damaging the eels. Instead, the crowds would have to be content with the sight of an electrical spark drawn from the eel daily at one o' clock only. Not one of the eels survived the London Season of 1776–7. Yet, the eels would live on in a sort of cultural afterlife.[3]

As already mentioned, John Hunter had dissected one of the eels that arrived dead in Britain in 1775 on Baker's earlier failed attempt to bring live specimens across the Atlantic. Hunter presented his findings to the Royal Society in the *Philosophical*

Transactions. Hunter's close association with the electric eel gave rise to the dubious honour of his being named in an erotic poem on the subject of the eels. The 1777 poem was addressed to 'Mr John Hunter, surgeon' and was an eroticized exploration of the electric eel and a satire of the sexual habits of the elite. The reader of the poem was placed in the position of an observer to the eel and a lover engaging in sexual acts. In one lewd scene a 'wanton Dame' has all her 'wants supplied' by an eel in Piccadilly. She 'covets every spark' the eel gives her, 'provokes him', craving a touch of his 'electric fire'. One line imagined an eel smuggled under petticoats. The 'tempting form' of the eel warmed female hands until 'his length erected stands' and a pseudo-orgasmic crescendo gave her 'the electric stroke'. The poet who penned this ode to the electric eel was far from bashful about penile metaphors.

Other poems too imagined women rushing in hordes to Piccadilly to receive an electric spark from Baker's eels. In one poem the electric fire of the eel warmed the hearts of women, and in the following year several erotic poems were published that cast the eel in a suggestive light. In the eroticized eulogies to the electric eel, supposed letters from ladies of fashion who had been the eel's lovers, the author posed as a woman called Lucretia Lovejoy. Georgian erotica was often especially more gratifying to readers if a woman had, ostensibly, at least, penned it. One of these was somewhat irreverently dedicated to the honourable members of the Royal Society who were at that very time engaged in experimenting on the eels. The sexual pleasures of the eels were to be had at a much lower price than those available in different parts of town. The association between electricity and sexuality in the Georgian imagination made it possible, in the realm of humorous erotica, for the spark given by the eel to be imagined as a sexual experience. In fact, in common English parlance at that time the 'spark' itself could easily be imagined as erotic: a 'spark' was defined as a very 'small part of fire' as well as a 'brisk gallant lover'.[4]

The cold weather of London in winter had a soporific effect on the eels and some fell into a state of torpor before they died. Indeed, there was only one eel left alive by early 1777. George Baker's loss was turned to comedic effect by the writer James Perry. He imagined the warm hands of London's society women attempting to revive 'their mutual friend' in vain as he died. After all, no eel could 'flash forever'. Enamoured women denied their spark would turn away in disgust: 'ladies too will turn their eyes and deign the thing to feel'. As exciting, or indeed erotic, as eels were in life, nobody wanted to stroke a dead eel.

There was a strong association between the readership of erotica and participation in Baker's electric eel spectacle. His exhibition was critical in presenting these living eels to Georgian London, and all of these erotic works mention Baker's exhibition in some way. This small, and admittedly rather strange, genre of electric eel erotic satire would not have been published were it not for Baker's demonstration of the 'electrical spark'.

In the eighteenth century pornography and erotica were not necessarily solely enjoyed in furtive privacy; many such Georgian works were worthy of sharing and laughing at over a glass or port. The sharing of erotica was often part of the convivial atmosphere of men's dining clubs. Periodicals such as the *Covent Garden Magazine* (1772–5) were bursting with salacious 'whore biographies' and engravings; other works like the novel *The Secret History of Pandora's Box* (1742) were slightly more cerebral, steeped in classical allegories. Other Georgian erotic material is rather baffling; *The Natural History of the Frutex Vulvaria* (1747) was crammed with botanical riddles along with more mundane erotic material. Written for sharing, electric eel erotica percolated through Georgian London in 1776 and 1777.[5]

Grasping an electric eel in person, or witnessing the electric spark in a dark room, was a privileged experience limited to the wealthy few Londoners able to pay for a spark; and this experience was fetishized for an elite male readership. Baker's exhibition displayed in a novel way the existence of animal electricity, but also made visible the 'electric spark' that became suffused with erotic meaning. As Georgian Londoners grasped the heads and tails of electric eels they would have thought about the erotic connotations of the eel, and at least some of the masculine elite had read the electric eel erotica then in circulation. In Baker's darkened apartment on Piccadilly, the sparks of electricity coursed through the whalebone corsets of women as they took their 'electric stroke' and illuminated the room with strange light. A reading of the Georgian electric eel erotica intended to entertain and titillate suggests at the jocular knowing looks that might have been exchanged between certain gentlemen at Baker's exhibition, or the private thoughts discreetly kept to oneself.

Exotic animals on display were conspicuous sources of satire and humour, and in a print culture dominated by a male elite, women were often the target of this humour. In the 1760s the 18-year-old Queen Charlotte was the esteemed subject of a series of jokes that revolved around 'Her Majesty's ass'. The crowds that gathered to see Queen Charlotte's zebra observed this animal with an assortment of sly risqué quips and *bons mots* in mind.

CHAPTER 15

The Queen's Ass

IN SEPTEMBER 1761 KING George III married the 17-year-old Duchess Sophia Charlotte of Mecklenburg-Strelitz. A somewhat belated wedding present arrived the following year, in July 1762 aboard HMS *Terpsichore*. Sir Thomas Adams, the captain of the ship, had wanted to give a pair of zebras to the new Queen, but only the female made it to England alive. This zebra, and later another, became celebrities, and both Charlotte and her eldest son George became strongly associated with the 'Queen's Ass'. These zebras became a symbolic representation of the royals and were used to criticize the Hanoverian royal family. In a letter to Jean-Jacques Rousseau, Voltaire warned him about the tendencies of the English news writers and their habit of keeping track of popular jests and the misdemeanours of the elite. Voltaire told Rousseau that he would be gossiped about, 'as they do the Queen's zebra, the English love to amuse themselves with oddities of every kind but this pleasure never amounts to esteem.' He wrote this in 1766; it had been only four years since the first of Queen Charlotte's zebras arrived in London. Perhaps even Voltaire would have been baffled by the enduring nature of the 'Queen's Ass' in Georgian culture. The print press and public derived a good deal of pleasure and amusement from the Queen's zebra, as zebra jokes were well worth telling for decades.[1]

The court of King George III and Queen Charlotte was not known for excess and splendour. Critics at home and abroad ruthlessly characterized the royal circle as parsimonious and unfashionable. George's simple tastes led him to become known as 'Farmer George',

8 *Zebra*, 1763, George Stubbs (1724–1806), oil on canvas

and the happy domesticity of the royals' private life set for some a moral exemplar. At first, the arrival of the young Queen's zebra prompted a flurry of sexual innuendo, but this flagged as it became unlikely that she would offer anyone a look at 'Her Majesty's Ass'. Instead the zebra became a symbol of political rivalry and corruption. In Georgian London aristocratic life did not revolve around the staid court as much as it did around the city's grand residences like Devonshire House or fashionable squares like St James's. The beau monde, and especially the Whigs, constituted their own envied circle as arbiters of taste. Prince George's residence, Carlton House, became one of late Georgian London's most fashionable addresses; the excessive spending and sexual indiscretions of the Prince supplied London's caricaturists with plenty of grist for their mill.

The two zebras that came into the Queen's possession attracted large crowds to their stable at Buckingham Gate. The first zebra that arrived in 1762 could be seen feeding in a paddock near Buckingham House as a gratuity of the Queen. A painting of the zebra was made from life and hung in the mews stables for those who could not get close enough for a better view. This zebra was by no means docile; a keeper needed to warn visitors to avoid a kick when they approached her. Although entry to see the zebra was supposed to be free, the Queen's Guard intended to make a little extra from exposing the Queen's Ass to public view. Sooner or later the petty pilfering caught the attention of the newspapers and the ensuing public outcry in 1764 caused the Queen's regiment to order the guard to refrain from such 'unbecoming practices in the future'. Even this order and the dismissal of servants who had demanded and taken money proved ineffective. London newspapers continued to print indignant letters from a public that fumed at the guards who 'have absolutely refused to show the zebra without being paid for it'. Some commented that these 'petit practices' did not do credit to the Queen, and that it was a wretched state of affairs indeed when such behaviour took place under the 'very eye of Majesty'. Some cynically supposed that the Queen did know of this and was using the zebra for her own financial gain. The zebra was drawing in scores of visitors from home and abroad: 'What must foreigners, who judge the whole by the characteristics of the few, think of such sordid doings?' The draining of the public's purses with apparent impunity offended a sense of honour and national pride. To add insult to injury, after paying a three-penny admission fee to those 'paltry wretches' who demanded them, the crowds who crammed into the stable or around the zebra's paddock then made a tempting target for pickpockets; thus the Queen's Guard and thieves were both robbing those hosted at the Queen's pleasure. The taking of monies persisted unabated; when the haberdasher Richard Hall and his family went to see the Queen's zebra on 1 April 1768 they paid two shillings for the pleasure.[2]

The Queen's first zebra had only been in the city a month when London's satirists churned out a slew of ass-related songs, ballads and satires. In August 1762 the crude yet imaginatively named 'FART-inando' the 'ASS-trologer' published a song called 'The Asses of Great

Britain'. Other rump-related humour was bawdier and attempted to get laughs at the expense of Queen Charlotte. As a young woman she was an easy target for London's more salacious satirists. Her zebra in particular inspired a tediously lengthy ten-verse allegorical song, the highlight of which was: 'What prospect so charming!/What can surpass?/The delicate sight of her M————'s A——?'[3]

It is clear that Georgians loved a little bit of semantic horseplay, and 'ass' had a wide range of associations, connotations and double entendres to work with. Depending on pronunciation 'ass' and 'arse' were also sometimes homophones. Georgian idiomatic expressions made much use of the word 'arse': to 'hang an arse' was to be lazy or tardy, a clumsy person might trip over 'arse-versy', and the more aggressively vexed might make use of the retort 'Ask my arse!' Not all the humour written to dubiously honour the arrival of the zebra was mean spirited or bawdy. Some of the humour was intended to charm, like the tale of the zebra arriving in London, allegedly written by a Frenchman. The zebra was written as if she were a fine noble lady arriving at the palace, with a guard placed at the door of her stable; regrettably her apartments were not fitted up to receive much company, so her picture was drawn to satisfy the crowds who could not have the happiness of approaching her in person at the 'Asinine Palace'. The writer playfully supposed that this 'charming beast' with her 'shining tabby skin' was a beauty to rival the Duchess of Modena, who would surely not have been received with such marks of distinction as Queen Charlotte's zebra.

Financial criticisms could take other forms too, and the conspicuous monochrome equid at court became an easy and amusing target. In the 1780s the Queen's second zebra was used to mock the vain character and spendthrift habits of the Prince of Wales. In 1787 a caricature, 'The Queen's Ass', was made of him, dressed in a zebra-striped three-piece-suit. Amusing as it was, the zebra suit also served to undermine George's gentlemanly taste and Englishness. In eighteenth-century politics simple masculinity was often equated with political virtue. It was the clothes that made the man, and if the Prince wore a zebra suit then he was, of course, an ass. The caricature's publisher, no doubt a little worried at the offence the image might cause, decided to omit 'Ass', and instead drew a painting of a braying zebra on the wall. This sartorial satire was certainly salient, as later that year the poet William Wallbeck penned a short poem on the matter. His 'Zebra and the Horse' poked fun at the Prince in his 'fine coat', suggesting that he resembled so perfectly the 'zebra insolent and proud, kept in the King's menagerie'. The showy over-spending of the Prince was a stark contrast to the penny-pinching of his parents. Much earlier, in February 1763 at the end of the Seven Years War, the cost-cutting absence of any fireworks or celebrations following the Peace of Paris prompted a tongue-in-cheek suggestion that the Queen's zebra ought to take part in a mock-battle instead.

The humour to be garnered from the mere hint of an allusion to the Queen's Ass is found too in letters between the Rev. William Mason and Horace Walpole in 1773, some ten years after the arrival of the Queen's first zebra. Since both the clergyman and the aristocrat were aesthetes, their ass humour was a touch more highbrow than the usual zebra satires. The wit of Mason's letter was predicated on Walpole already being absolutely au fait with the jokes that circulated about the Queen's zebra. Mason wrote to Walpole with evident pleasure, dressing his witty titbit as a dull letter from provincial Yorkshire:

This dull place affords me no news except that her Majesty's zebra, who, according to the advertisement in our York Courant of this day, it seems was lately the property of Mr. Pinchy and purchased by him of one of her domestics (although, as I rather suspect, given to him for the valuable consideration of his friendship) died the third day of April last at Long Billington near Newark. This advertisement further adds 'that the proprietor has caused her skin to be stuffed, and that upon the whole the outward structure being so well executed, she is as well if not better to be seen now than when alive, as she was so vicious as not to suffer any stranger to come near her, and the curious may now have a close inspection, which could not be obtained before'. She is at present exhibited at the Blue Boar in this city with an oriental tiger, a magnanimous lion, a miraculous porcupine, a beautiful leopard and a voracious panther, etc., etc.

The 'Pinchy' who had put the zebra on display was Mr Christopher 'Pinchy' Pinchbeck, clockmaker and friend to King George III. Pinchbeck had perhaps been given the zebra as a favour, though he may well have bought it. Although the zebra was no longer on display at Buckingham Gate she was still strongly associated with Queen Charlotte. The zebra made a distinct impression on the people of Oxford and fellows of the university, as reported in *Jackson's Oxford Journal*: 'of all the natural curiosities exhibited at this university, nothing ever drew the attention of the curious so much as the beautiful and astonishing zebra lately belonging to Her Majesty and generally called "The Queen's Ass".' This joke was certainly one that died hard.[4]

When the zebra eventually died she was stuffed and then put on display at the Blue Boar Inn in York. This was a great deal less salubrious than her accommodations at Buckingham Gate. Her temper and tendency to kick spectators had prevented close inspection in life, but in death one could get a far better view. After sketching out for Walpole the course of events, Mason with a literary flourish got down to the juicy stuff and took a shot at the Queen:

> Pray do you not think the fate of this animal truly pitiable? Who after having, as the advertisement says, 'belonged to her Majesty full ten years,' should not only be exposed to the close inspection of every stable boy in the kingdom, but her immoralities whilst alive thus severely stigmatized in a country newspaper. I should think this anecdote might furnish the author of Heroic Epistles with a series of moral reflections which might end with the following pathetic couplet: Ah beauteous beast! Thy cruel fate evinces, How vain the ass that puts its trust in Princes![5]

Mason might have somewhat over-egged the pudding with his allusion to Ovid's *Epistles*, but he made his point. The bad-tempered zebra had her 'immoralities' stigmatized in a provincial newspaper and was exposed to the 'close inspection' of stableboys; a turn of phrase with sexual undertones. From Buckingham Gate to the Blue Boar – such were the cruel vicissitudes of life. The afterlife of Queen Charlotte's second zebra was more dignified. When it died, this zebra was presented to the Leverian Museum, where the relationship between the Queen and her zebra was noted in guides to the collection.

Georgian naturalists hoped that zebras might possess a nature malleable enough to render them useful. But the personalities of the Queen's two zebras did not offer much encouragement. Both were known for their biting, kicking and 'ungovernable behaviour'. The Queen's disposal of her zebra might have been motivated by a desire to put an end to the ass jokes. But it was probably also motivated by ennui and the zebra's attitude. Confinement and a diet supplemented with tobacco can have done little to calm the animal. Queen Charlotte might have hoped that her second zebra would be better behaved. It was not, and this zebra was sent to the Tower. Some, like the Queen's biographer, attributed this to the 'rudeness of the populace'; crowds of people might have spoilt the regal ambiance of Buckingham House, though perhaps by 'rudeness' the biographer was alluding to the 'Queen's Ass' rump humour. The irritability of her second zebra was such that in her new home, the Tower of London, she grabbed her keeper with her teeth and threw him to the ground. He might have been trampled if he had been unable to beat a hasty retreat. This sort of behaviour led naturalists to conclude that the wild and independent zebra was 'ill adapted to servitude and restraint'. One notable exception had been the male zebra at Exeter 'Change, before he died in a fire. This zebra, unlike his kin, 'appeared to have entirely lost his native wildness' and was so 'gentle as to suffer children to sit quietly on his back, without any symptom of displeasure'. William Nicholson thought that there was 'little hope' that the 'vicious' zebra could be of 'great service to mankind'. However, he did indulge in some wishful thinking: if the zebra could be used in the same way as the horse, 'an elegance and variety would be added to the luxuries of the great and the opulent.' Nicholson's Enlightenment dream of zebras pulling the carriages and phaetons of the rich around London was not realized, but it was not too far-fetched an idea for Georgians in the 1790s. After all, a colony of kangaroos belonging to the Queen grazed on the banks of the Thames at Kew, with others hopping around on the estates of aristocracy and gentry.[6]

Queen Charlotte's zebras were not the only ones to be found with royal associations in Georgian Britain. Fredrick Louis (1717–51), the Prince of Wales, had kept both a male and female zebra at Kew in the 1740s and 1750s. Later, in 1780, a zebra arrived in London that had been taken as booty from a Spanish ship; it had been intended for the menagerie of King Charles III of Spain. Never to graze under clear blue Iberian skies, the zebra found itself under the grey London clouds instead. The erstwhile royal zebra was exhibited at the Bell Inn on the Haymarket for one shilling, and eventually advertised for sale for the huge sum of 400 guineas, or £420. The zebra as a royal mascot permeated Georgian culture in other ways too; for example, in the naming of two naval warships. The first HMS *Zebra* was launched in 1777, but it was in service for only a year before it was scuppered and blown up during the American Revolutionary War. A successor, a second HMS *Zebra* was launched in 1780.

Queen Charlotte's zebra, or rather the Queen's Ass, lingered on in Georgian culture life, not only in saucy humour or political satire, but also in more serious contemplations on the nature of the zebra and adaptability. When Georgian Londoners went to Buckingham Gate and grudgingly coughed up their money to see the zebra, many would have been familiar with the satirical prints in circulation. For those Georgians who remembered the zebra of the 1760s and that of the 1780s, the zebras came to represent the Queen. The lives of the zebras and their Queen became entwined. In his biography of the Queen, John Watkins presented a sugared, almost hagiographical, account of her life, a life of domestic happiness and patronage of charitable institutions. On the matter of the Queen's zebras, however, Watkins allowed himself a mere flustered few lines. He had not even been alive when either zebra had been in London, but he certainly knew of them and the extent to which the English, as Voltaire had said, loved to amuse themselves with oddities of every kind. Watkins might have furnished his biography with at least a few of the less crass jokes. Instead he noted only that 'a female zebra attracted much notice and excited considerable amusement' and then added the rather tame pretty little story about the zebra, allegedly written by a Frenchman – the story in which the irreverent humour peaked at describing Buckingham Gate as the 'Asinine Palace'. Watkins need not have been so circumspect; even the most sincere Georgian readers of his regal biography might have permitted themselves a smirk or two as they paused to recall the 'considerable amusement' the zebra had excited. After all, jokes about the Queen's Ass had been worth telling for decades.[7]

The zebra, like the electric eel, became an animal used to convey and articulate a wide range of cultural concerns in the late eighteenth century, as well as inspiration for humour and satire. These particular animals were highly conspicuous exotic oddities in Britain in this period, but far more common exotic animals also attracted attention. Many women from the elite and 'middling sort' kept parrots as exotic luxuries. With privileged access to the private spaces of the boudoir or parlour and with physical intimacy with their mistress, these birds troubled appropriate ideals of relations between the sexes. In print culture, the parrot became imagined as a rival suitor in the game of love, or as a wretched bird for which a woman squandered her life.

CHAPTER 16

The Love Birds

IN 1709 TATLER PUBLISHED a short poem, 'To a Lady on her Parrot', by Richard Steele. Isaac Bickerstaff, Steele's alias, had supposedly visited a London coffee house and met a heartbroken man whose enamoured sighs went unnoticed by a 'fine lady'. As he languished, the lovelorn man penned a heartfelt poem to this lady's parrot, hoping earnestly to convince the bird to put in a good word for him with its mistress. The parrot was imagined as both a rival suitor and the incarnation of the poet's own carnal desires. The parrot-seducer 'swells his glad plumes' in an 'amorous trance'; Steele even proposed that 'the parrot represents us in the state of making love' and that 'henceforth the parrot be the bird of love'. This poem was reprinted throughout the eighteenth century and inspired others of an even more erotic tone. These sorts of poems were meant to be humorous, especially perhaps for men since this humour is generally at the expense of women. The ownership of parrots by women became something of a controversy, with surprisingly staunch opponents. Parrots as the recipients of their mistresses' caresses offended certain sensibilities; moreover the 'prattle' of a parrot became increasingly seen as akin to the idle and vacuous conversation of women. As such the parrot represents Georgian attitudes towards women, sexuality and women's conversation.

In Georgian children's poems, songs and stories parrots are the playful companions of young ladies; they perch on hands, receive strokes, give a kiss to their mistress and crack nuts fed to them lovingly. The rhyming verse 'Miss and her Parrot' (1751) was a jolly song in which a little 'Miss' calls upon her 'Poll'; 'Perch upon my hand, take a while thy stand', 'Pray give a kiss to thy own dear Miss', 'I will stroke thy back, give thee nuts to crack'. The practice of keeping native songbirds was long established by the seventeenth century, and later canaries too became familiar cagebirds. By the mid-eighteenth century canaries were common enough that the children of labourers and artisans could reasonably hope to own one. Parrots, however, remained expensive and a conspicuous status symbol of the mercantile and upper classes; macaws, for example, sold in London during the 1730s and 1740s for ten guineas, about the annual wage of a domestic servant.[1]

In adulthood, the ownership of a parrot had a different set of connotations. The portrait of Susanna Duncombe, née Highmore, painted by her London-based artist father Joseph Highmore, features a small green redheaded parrot clambering on a gold ring, perched perilously above two tortoiseshell cats. The young Susanna, on the cusp of womanhood, clasps an oval portrait of a boy, an allusion to young love. Her parents were from moneyed mercantile families, and her father enjoyed patronage as an accomplished portrait painter; as such, like many young women of her position, Susanna came to own a parrot. Susanna was admired for her beauty and intellect in London's artistic and literary circles, and her father noted that in the same way she had carelessly burnt herself with her curling irons, so too she often set the hearts of young fellows on fire'.

Highmore would link his daughter with her parrot again when a poem of his was published in the *Gentleman's Magazine* in 1750, when his daughter would have been around 20 years old. Apparently Susanna, 'desiring her parrot might be the subject of a poem', asked her father to pen one in his honour since Poll in his 'gilded prison hung sung to all' but was 'left unsung'. Her father imagined Poll as a 'saucy rogue' who might 'woo as lovers do' but in the case of Susanna he hoped the feathered rogue would guard his daughter and let no smooth-talking 'fopling wit' go admired for his repartee; since they were, no doubt, themselves taught by a parrot to utter sweet nothings by rote.[2]

In wider print culture the relationship between women and parrots was a subject of satire and censure. Here parrots are held close to their mistress's breast and peck lovingly at her hair and neck; the parrot is imagined as a substitute lover or rival suitor, not an appropriate childhood companion.

9 *Susanna Highmore*, c.1740–5, Joseph Highmore, oil on canvas

In a chapter with the dour title 'Of Adversities and Afflictions', *The Young Gentleman and Lady Instructed in Such Principles of Politeness* (1747) helpfully instructed its purportedly genteel reader on the nature of grief, sorrow and tears. Indeed, though 'many are the children of affliction' there were at least a few sorrows that were felt especially by women. In particular 'the death of a parrot' was one of the more distressing misfortunes to blot the life of a young lady; 'neglect at a ball or an assembly' or 'being left out a masquerade' being some of the other hurtful 'afflictions' among the fairer sex. Tears wrung for 'pretty Poll' were for some a touching example of feminine sensibility but for others suggested an indulgent and excessive tenderness. In 1778 a reader of the *Lady's Magazine* from Bedford wrote in to admonish women for the affection they felt for their parrots. Such love was 'ill-directed tenderness and sensibility' and when a parrot became the 'bosom companion' of its mistress she might even abandon her own offspring to a servant. Such compassion was misdirected, in the same way that a woman ought to caress her lover, not her bird. The writer claimed, 'I have known many a fair one all bathed in tears for the loss of a favourite parrot,' even if she was hardened to the plight of the poor, orphans or distressed widows. The love of a parrot could be a selfish love that distracted a lady from her proper duties. 'Miss Rattle' – a mere consonant away from Miss Prattle – sits alone at her breakfast table in a caricature from the 1770s, the epitome of a lady whose misdirected sympathies cause her to neglect her womanly duties and so end up alone.[3] In a later caricature, a lady sits alone with her 'family': a veritable menagerie of monkeys, parrots, dogs, cats and a magpie.

Instead of a parrot receiving the caresses better deserved by a gentleman suitor, at least one parrot joke sought to entertain the idea of a beloved parrot providing unwitting connubial respite in a marriage. Mrs Ginger, the wife of a City of London alderman, had a higher sex drive than her husband and disturbed him with her 'nocturnal hints and persecutions'. The wily alderman knew that 'one cross word to the parrot' and his wife would leave him, foiled in her attempts to initiate sex, to 'sleep in laziness', refusing to share a bed

A MAIDEN LADY AND HER FAMILY.

10 *A Maiden Lady and her Family*, c.1820–40, Orlando Hodgson, watercolour lithograph

with him. So the joke went, the alderman said to his wife, 'Damn your parrot, he's as hoarse as a raven!' And the man had his bed to himself ever since. The snoring alderman's sexual reluctance was not the only source of laughter here; his sexually frustrated and scolding wife's love for her parrot meant the erstwhile 'bird of love' guaranteed him a good night's sleep. Parrot-based humour at the expense of women also took a different form, that of eulogies. Eighteenth-century elegies and epitaphs to the death of a favourite animal were a literary form that either expressed or mocked refined sensibilities, as well as providing an excuse for the writers to show off their literary virtuosity. Often these elegies and epitaphs made light of the grief that had been called by others an 'affliction'. 'To a Young Lady on the Death of a Favourite Parrot' (1773) supposed a grief so deep that, naturally, only the love of another – and presumably a man – could soothe the pain: 'her heart with wasting anguish burns, which some fond heart must quell'. A rather earthier epitaph, 'On the Death of a Favourite Parrot', written in 1782, commemorated the ignominious burial of a parrot in a necessary house. Poor Betty's parrot had been thrown down the privy chute to spare the expense of a more decent interment. With tears in her eyes, his weeping mistress promised never to visit without 'leaving something behind her'.[4]

A parrot was not always a corrupting recipient of 'ill-directed tenderness'; sometimes it was this tenderness that might preserve a woman's honour. A parrot could tame the passions or at least provide a safe outlet for them. One writer supposed that if a woman was 'exposed to the unruliness' of romantic and sexual desires these 'tendencies' would be removed by 'substituting' some object to 'engross her attention'. A parrot might relieve her distress if it was 'admitted to her bosom and received those caresses which her passion prompts her to put upon her lover'. Here a feathered lover was more suitable, and for modesty's sake more desirable, than one in a coat and breeches.

The tone of the poems that imagined a parrot as an ersatz lover is captured in 'The Lover and his Parrot' (1732). Here a lovelorn man orders his parrot to 'swift to my cruel mistress fly'. He hopes the parrot will win her over and in being intimate realize his own desires. The erotic overtones probably spared few blushes: 'with thy bill her cheeks caress', 'tread softly over her bosom's charms', and 'flutter on her snowy arms'. Feathered tokens of affection were not simply a Georgian literary conceit; a parrot, and especially one trained to repeat its new mistress's name, put other trinkets to shame. Moreover, since the parrot was the bird of love in Georgian print culture, both recipient and giver were knowingly gesturing towards intimacy.[5]

Ann Sheldon's life took many guises: lover, mistress, prostitute and destitute. She could, for a while, count titled men as her suitors, patrons and benefactors. Yet, she too might just as easily find herself suddenly jilted, cold shouldered or turned out of her chambers in a virtually penniless state. In her situation, which was really quite perilous, she found that a former lover might so very easily turn from a charming suitor into a vindictive foe. Given this – and the fact that she was almost perennially on the run from her debtors in the 1770s – Sheldon was not unacquainted with the services of a pawnbroker. In hard times she had been compelled to dispose of her jewels and fine lace this way before. The day she came to find her 'favourite bird' – a 'beautiful and talkative parrot' –in the window of a pawnshop represented, however, a particularly vexing and embarrassing affair.

A certain Sir Alexander Gilmour, with whom she had shared some form of intimacy, had at some point given her a green parrot which he had assiduously taught to screech 'Miss Sheldon'. The parrot was something of a talker, albeit with a somewhat limited vocabulary of a mere two words, and so Sheldon's room would echo with squawks of 'Miss Sheldon, Miss Sheldon, Miss Sheldon' for a full five minutes.

One evening, Sheldon accompanied some army men to a bawdy comedy playing in Wandsworth. She went with her friend and protégé Miss Sally Metham. One of the men in this circle, Major Atkinson, was a regular acquaintance and patron. During a private performance, the assembled company found the comedians to be 'imperfect in their wardrobe' and so they 'disencumbered' themselves 'of a good part of our clothes' in order to more properly dress them. After the play had finished both women decided to leave the army men to their drink and accompanied the comedians to an 'excellent supper'. In lieu of paying for the supper in cash, the women left the army gentlemen's clothes in pawn. Major Atkinson and company, on sobering a little, were forced to borrow garments from the master of the inn and hunt down the troop of comedians to recover their rightful property. Major Atkinson would have his revenge.

Atkinson knew, obviously, of the precious green parrot that noisily occupied Sheldon's room (a rented residence on Delahay Street, Westminster) and enquired if her parrot might teach his own parrot 'the art of speech'. The thought of an 'additional serenade' to her own parrot's unremitting 'Miss Sheldon'-ing was not an agreeable one, so Sheldon agreed that he could borrow her bird. As it turned out, this psittacine language exchange did not happen. Atkinson did not in fact own a parrot and instead, along with the purloined bird, headed straight to a pawnbroker's in Princes Street, Leicester Fields. Atkinson ordered the bird to be displayed in the shop window – he wanted this little bird to be seen and heard.

Naturally, passersby were treated to a barrage of 'Miss Sheldon' and it was not long before some of Sheldon's acquaintances, Sir Cecil Bishop and a Mr Sweedland, passed by. They headed immediately to see Miss Sheldon and rib her for pawning her parrot. She immediately sent to know what sum the pawnbroker wanted for the parrot, but he would not return it. Major Atkinson had ordered the man not to release the bird. Sheldon was forced to implore Atkinson to release her bird and bribe him with a bottle of claret. He agreed but would only return the parrot under the cover of dark. His reasoning for this was a spiteful joke; the parrot ought not to see the way from the pawnbroker to Miss Sheldon's, since it was not fitting for the bird to look upon the quarter in which his mistress ran her 'cursed errands'.[6]

Far away from a pawnbroker in Leicester Fields or the chambers of Ann Sheldon, parrots could be seen in London townhouses and country seats. Bulstrode Park in Buckinghamshire, the seat of the Duchess of Portland, Margaret Cavendish Bentinck, was well known for its extensive art and natural history collections. The letters of the Duchess's friend, Mary Delaney, who often stayed at Bulstrode, describe a house teeming with birds. Delaney wrote to her friend, a Miss Dewes, in September 1768: 'We have had just breakfasted, and the little jonquil parrot with us; it is the prettiest good-humoured little creature I ever saw.' Like other aristocratic and fashionable women, the Duchess took breakfast in her bedroom or dressing room instead of a breakfast room. A lady's morning toilette was not necessarily a private affair; instead several hours might pass as guests visited and took tea as the mistress leisurely dressed, applied cosmetics and perhaps stayed in a morning gown until the early afternoon, reading or writing letters. In the most aristocratic of households this morning ritual was called a levee and might be a crowded ceremonial occasion. Parrots were often kept in these sorts of rooms and are pictured perched on stands or swinging in cages. As such, parrots would have been a familiar part of elite life.

Delaney's letters also praised the Duchess's collection: 'such beauties of foreign birds', 'her birds are so many and beautiful', and these were just the birds kept in cages inside the opulent grand house. In the elegant parkland a whole host of 'macaws, parrots, and all sorts of birds' squawked and fluttered in their aviary. Delaney's letters mention the birds of other aristocratic women too, including the Duchess of Norfolk, though she 'only saw a crown bird and a most delightful cockatoo' in her menagerie. The amorous parrot with his 'swollen plumes' was a bird at the heart of feminine intimate and social life, and so it was a good candidate for literary or poetic satire.[7]

'Our military operations are at last begun. The heart of every Englishman now swells with confidence.' So wrote Samuel Johnson on 13 May 1758. The Anglo-French conflict in North America had extended into a continental war in Europe, and now red-coated soldiers were camped on the Isle of Wight awaiting dispatch overseas. Johnson was writing his weekly essay known as the 'Idler' and published in the London weekly *Universal Chronicle*. Johnson's essays were known in the city's lively print culture and he seemed to relish pleasing his readers with his satirical and irreverent take on affairs. With the 'men of scarlet' away, Johnson's mind turned to the plight of London's women: 'How shall the ladies endure without them?' With the city's young gentlemen abroad London's society women were 'left to languish in distress' in the absence of male company. A lady must now sit at a play without a critic, choose a fan by her own judgement, walk in the Mall without a gallant and shuffle cards 'in vain impatience for want of a fourth to complete the card party'. With a dash of tongue-in-cheek pity Johnson hoped that these women had at least lapdogs or monkeys to keep them company, unsatisfactory companions though they were. Of course, a parrot made for a far better companion than a dog or a monkey for a woman languishing in the absence of a man. Johnson, known for his wit and not least his *Dictionary of the English Language*, chose his words prudently and with intent. Naturally, he knew that his readers would chuckle at the thought of parrots substituting for absent gentlemen. Indeed, in his dictionary the parrot had been noted as a bird particularly remarkable for 'its exact imitation of the human voice'. But Johnson took his satire further still and deftly drew a parallel between polite society conversation and the prattle of a parrot: 'a parrot is indeed as fine as a colonel and if he has been much used to good company is not wholly without conversation.' The loquacious parrot need not leave London's society women conversationally bereft. But, at the end of the day, the parrot was after all a 'poor creature'; he has 'neither sword not shoulder knot and can neither dance nor play at cards'. The parrot was not a dashing and attentive suitor, but he was at least better than none.[8]

Some years later, on 19 June 1775, Johnson wrote to his friend Mrs Hester Thrale, or Hester Lynch Piozzi, to tell her that 'Mrs. Aston's parrot pecked my leg, and I heard of it sometime after at Mrs. Cobb's.' Johnson's circle probably found it amusing that Poll had his revenge, and enjoyed a little gossip.[9]

Samuel Richardson is famed for his novels *Clarissa* (1748) and *Pamela* (1740). These are epistolary novels – narratives written in the form of letters – that deal with matters of love, morality and sexuality. These books were popular, even notorious, in the Georgian period, though Richardson's last epistolary novel is not considered his best. *The History of Sir Charles Grandison* (1753) was panned by contemporaries and is largely forgotten as a folly of Richardson's autumn years, overshadowed by its racier and celebrated sisters *Clarissa* and *Pamela*. Sir Charles's love for the heiress ingénue Harriet Byron is ultimately thwarted as he marries and indeed loves an Italian noblewoman. As a platonic friendship rather than a consummated romance blossomed between Charles and Harriet, so too did a deep friendship form between Harriet Byron and Charlotte Grandison, Charles's sister. In the chain of letters that construct the narrative of this novel a parrot appears several times; a literary and imaginative touch of Richardson's intended to play upon his reader's minds or elicit a smile. In these letters some familiar eighteenth-century attitudes towards parrots and women emerge.

On 4 August Charlotte writes to Harriet with some 'pen prattle', in particular she wished to divulge an anecdote concerning the charming devotion of her brother. She wrote, 'God bless his soul, he came to me now so prim and pleased.' Charles had given her a parrot and a parakeet. Her brother 'had great difficulty' in getting them and the parrot was 'the finest talker!' Charlotte told Harriet that she would not suffer any other woman to impose so on her brother: 'I will not allow anybody but myself to abuse him.' Richardson's description of Charlotte's writing as 'pen prattle' alludes to the prattle of a parrot. And a young man too eager to please women with expensive toys is abused and put upon. Charlotte was not, however, the only woman in the household with a parrot. Their maiden aunt too had a parrot, and in a different letter Charlotte was particularly cruel. Scarcely a month after her brother had given her her own parrot, Charlotte wrote again to Harriet. Her aunt was being bothersome and fussing over Charles, who had recently returned home: 'I always think that when I see these badgerly virgins fond of a parrot, monkey, or lapdog, that their imagination makes out a husband and child in the animals.' Parrots could be love birds and substitute suitors given to women in their alluring youth. Or they might signify something more tragic – a failure in womanhood and an air of desperate spinsterhood. Richardson's 'badgerly virgin' and her parrot drew upon Georgian attitudes towards womanhood in a salient fashion for his readers. The last appearance of the parrot in the novel is when Charles Grandison speaks to Harriet and tells her that she will see many things 'worth your notice' in his collection. His sister is uninterested in the collection: 'nothing less than men or women' are worthy of her notice, with the exception of her parrot and squirrel – 'the one for it prattle, the other for its vivacity'. As alluring as a collection of preserved insects might be to a gentleman or bluestocking, Charlotte's mind is concerned only with trifles and prattle, not a natural history cabinet.

Etiquette and conduct guides throughout the eighteenth century admonished women for their prattling. *The Whole Duty of Woman* (1753) laid its advice on thick with 24 chapters, each titled according to the respective feminine state, vice or virtue upon which William Kendrick wished to expound: 'Vanity', 'Virginity', 'Frugality', 'Complacence' and 'Marriage' give a taste of the tone Kendrick set. In 'Reflection' Kendrick supposed that women ought to mind their words carefully, lest they rattle on mindlessly and dishonour themselves. Indeed, 'as the green parrot squalls without ceasing, so is a woman who regards not her speech'. The speech and conversation of women in a patriarchal society could then be imagined as ceaseless squawks. *The Polite Lady* (1769) too bemoaned the sort of women who were barely able to 'form a good thought' and instead merely related the 'trifling' and 'insignificant sayings' of others, like a 'prattling parrot'.[10]

Letters sent in 1744 between Elizabeth Montagu and her friend, the Rev. Freind, suggests that many women dreaded being called a 'parrot' and indeed that some women were quick to dub others parrots. Freind had expressed anxiety about his wife being 'taken as a parrot'. Montagu assured him that this was not the case; 'She has less of the parrot than any woman in the world; a parrot is the very reverse of her character; but I assure you I know a great many parrots; I met four in a visit yesterday'. Georgian men and women alike, then, took to using 'parrot' as shorthand for a prattler or highly talkative woman. Montagu hosted literary breakfasts and evening assemblies; as hostess she gathered together in her salon a coterie of Georgian London's prominent thinkers, writers and wits. As such, outside this esteemed circle, there may have been many whom Montagu found to be, comparatively speaking, dull or tedious 'parrots'. She playfully compared herself to a parrot in a different letter, this time to her friend Mrs Donellan. Montagu, also staying at Bulstrode, like Mary Delaney, told Donellan that she wished to spend a great deal of time together when they met, so she was to expect her, good friend that she was, sat in her dressing room 'as chatty as your parrot'.[11]

Tobias Smollett, novelist and poet, in his splenetic and caustic *Travels through France and Italy* (1766) gave his own scathing assessment of French male conversationalists. French men were, according to Smollett, from a young age engaged in frivolous pursuits, including becoming a 'connoisseur of hairdressing' and accomplished in polished conversation: 'like a parrot he learns by rote the whole circle of French compliments.' Moreover these suave French parrots put their prattle to use by throwing out their words indiscriminately 'to all women without distinction', so much so that Smollett considered a Frenchman's conversation 'no more than making love to every woman who will give him the hearing'. Well-preened loquacious French men, 'modesty utterly unknown among them', seem to have been something of a pet peeve of Smollett's, for he dedicated at least three pages of his book to the defamation of their character. Frenchmen appear as effeminate fops intruding into the sphere of women in a large number of eighteenth-century British caricatures. The women of the 'ton' or the beau monde were lampooned alongside their French hairdressers, dance masters and couturiers. English men of fashion too were often derided; in the case of the London 'beau', Smollett thought that 'like a parrot he chatters and struts like a crow'. Prattle was not necessarily the sole preserve of women, though they attracted the greater degree of censure. Although women were the target of satire and censure in popular culture, as the owners of collections of birds they could attain considerable status and recognition as patrons of natural history and ornithology.[12]

Martha Wager, née Earning, was the wife of Sir Charles Wager, First Lord of the Admiralty. Described as a 'curious lady' and 'great

11 *Guinea Parrot*, 1736, George Edwards (1694–1773),
watercolour, gouache and graphite

admirer of birds', Lady Wager accumulated a large collection of
living exotic birds at Parsons Green and Stanley House, Chelsea, in
the 1730s and 1740s. George Edwards, naturalist and librarian at
the Royal College of Physicians of London, described her collection

as 'a greater living collection of rare foreign birds than any other person in London'. She had procured these birds 'by presents and purchase'; her husband's naval career meant that officers eager to curry favour and seek promotion presented curious foreign birds to the Wagers. Lady Wager both paid Edwards to draw her birds and permitted him to come and draw them for his own publications. He noted in his *A Natural History of Birds* that it was only through her 'goodness in communicating to me the knowledge of everything new that came into her hands' that he was able to furnish his book with a collection of drawings from life and descriptions of the birds' origins and natural histories. A large number of birds came into her hands: at least six parrots and parakeets, some finches, doves and sparrows as well as Indian and African cranes. Her red-and-blue-headed parakeet from the West Indies was described as a 'beautiful little parrot' that chattered rather than talked, being a bird that spoke 'few words distinctly'.

Lady Wager as a patron was, then, a respected source of knowledge about exotic birds, parrots in particular. Her aviary was not a static collection, as birds came and went – presented to other individuals. In this way curious and foreign birds functioned as a means of forming and cementing social bonds. The Wagers' black parrot from Madagascar, though 'gentle' and 'always choosing to be on the hand', was given to the Duke of Richmond. Similarly, the ash- and red-coloured parrot that Edwards drew at the Wagers' house in 1736 was passed on to Sir Hans Sloane, and it was still alive at his house in Chelsea in 1750. Lady Wager's role as a patron was significant and she became a respected authority; however, outside the circle of London's naturalists and bird lovers she might have been perceived rather differently. The Wagers remained childless and later, as an aristocratic widow presiding over a collection of birds, Lady Wager was perhaps seen as a 'badgerly' dowager doting over a brood of parrots.[13]

Sarah Jodrell, Lady Ducie, or Mrs Robert Child, was one of the wealthiest women in late eighteenth-century Britain, having acquired a large degree of financial independence through her first marriage into the wealthy Child family. The family had made their fortune in the East India Company and the family's bank, Child & Co. Her collection of living birds at Osterley Park, then in the rural environs of London but now in the London Borough of Hounslow, numbered well over 100 species, acquired through private purchase and exchanged with friends, including her neighbour Sir Joseph Banks. Sarah Child and her husband Robert commissioned the artist William Hayes to write and illustrate a description of their collection. This volume was the first of its kind to include portraits and descriptions of an entire single collection of living birds. Earlier in the century Lady Martha Wager had paid George Edwards to draw her parrots but the project of the Childs was larger and more elaborate. Hayes's *Portraits of the Rare and Curious Birds with their Descriptions from the Menagery of Osterley Park* (1794–9) was a family affair; Hayes and his wife shared the illustrations, with the colouring done by their daughters Emily and Anne, and this volume presents birds, especially parrots, as the embodiment of femininity. Hayes's florid language characterized birds as graceful, beautiful, delicate and engaging; the birds seemingly display desirable feminine qualities.

The lesser white cockatoo was 'elegant', 'uncommonly graceful' and capable of 'deriving great pleasure from being caressed', and it thoroughly enjoyed 'being taken notice of', flying up to the top of the highest trees in the menagerie but always returning when he was called. The great red-crested cockatoo also received praise as possessing a:

> superior understanding to that of the common Parrot, and are more docile, kind, and sincere in their attachments. This amiable disposition was particularly manifested in the subject of this Plate, for its fondness, affectionate attention, and attachment to the person who had the care of it was beyond expression.

The cranes in the collection at Osterley Park were also pleasant and attractive; the Numidian crane:

is gentle and social, apparently much pleased at being admired, and embracing every opportunity of showing and setting itself off to the greatest advantage to those who seem attracted by its beauty; it accompanies visitors in their walk in the most graceful manner imaginable.

The crowned African crane was:

yet more gentle and familiar. It is much delighted with being taken notice of, and was a constant attendant to those who visited this delightful spot, making the tour of the menagerie, with slow but measured steps, and always parting the company with much apparent regret, which it expressed by raising the neck, and making a hoarse unpleasant cry.

The language of sensibility and the picturesque diffused into the assessment of exotic birds in other natural histories too, like Wood's *Zoography; or the Beauties of Nature Displayed* (1807). Here the cockatoo at the Exeter 'Change menagerie was feted as 'beautiful', 'picturesque', 'mannered' and 'attentive'. Notwithstanding this, the abominable screams of the cockatoo – like the hoarse cry of the crane at Osterley – resounded unpleasantly in the ears of admiring spectators. Sarah Child's collection, though held in great esteem during her lifetime and the subject of Hayes's *Portraits*, was dispersed shortly after her death. In 1795, the Rev. Daniel Lysons's *The Environs of London* with quite some regret informed its readers, 'we are sorry to find that the menagerie we have so frequently admired has been deprived of its beautiful feathered inhabitants since the death of Lady Ducie'. These beautiful feathered inhabitants had been, according to Horace Walpole, gathered from a thousand islands not even yet discovered by Joseph Banks. Almost certainly Sarah Child's collection of delicate and pleasing birds had been the largest of its sort in Georgian London.[14]

The humour to be had from electric eels, parrots, and the Queen's Ass indicates the degree to which exotic birds and animals permeated

12 *A Crane*, 1791, William Hayes (1729–99), watercolour,
gouache, ink and graphite

the cultural and material life of the city. *The Georgian Menagerie* has revealed the life of the city's animal merchants and bird sellers, and the significance that the birds and animals they sold held for those who owned and saw them. The diaries and letters of Londoners and visitors to the city, in addition to wills, insurance records, court records, newspapers and broadsheets reveal a city bursting with exotic birds and animals.

Epilogue

THE EARLY DECADES OF the nineteenth century saw a substantial change in the way exotic birds and animals could be seen in London and further afield. The rise of zoology as a discipline and the establishment of the zoological garden altered London's Georgian urban jungle. This transition from eighteenth-century menagerie to nineteenth-century zoological gardens is explored here through the story of the giraffe that came to London in 1827. The journey of this animal to London, his life and his afterlife can be used to trace the end of the Georgian menageries.

Saturday 11 August 1827 was an important day for Edward Cross. King George IV had made him responsible for unloading a present from the Ottoman Viceroy of Egypt, Mehmet Ali Pasha. But Cross had never even seen a giraffe before, and the male calf would be the first ever to arrive in England. As proprietor of the Exeter 'Change Menagerie, there were few as qualified as Cross to carry out the King's orders. The Pasha was hoping to dissuade the British from supporting Greek independence but his giraffe diplomacy was a failure, even as the giraffe was en route. A mere three months after the giraffe arrived in London, British ships would fire upon a Turkish–Egyptian fleet at the Battle of Navarino.

The giraffe had travelled a very long way after being caught by hunters in the Sudan. First he had been strapped to the back of a camel for over 40 days until he reached Cairo; from there he was shipped to spend the winter in Malta. An English ship, *Penelope*, took him aboard at Malta and arrived in London in August 1827. Cross had known for a few weeks that the giraffe would soon arrive on the Thames, and he had prepared a barge with awnings strung across the deck. This barge was brought up alongside the *Penelope* and the giraffe safely boarded with his entourage of two Arab keepers, an interpreter and two wet-nurse cows that provided him with milk. Around six o'clock in the evening the barge docked at the Duchy of Lancaster Wharf, London Bridge, and from here the giraffe was found lodgings in a warehouse. The crowd that had gathered on the wharf was kept out of the warehouse but the little giraffe, disturbed by the noise, kept looking anxiously towards the window.

On Monday a special caravan arrived, drawn by four horses, to convey him to George IV's menagerie at Sandpit Gate, Windsor. George apparently took a liking to his special gift and came to visit the giraffe two or three times a week. The public too, if they so wished and had the means to travel some 20 miles from London, could also view the giraffe. The King's private menagerie at Sandpit Gate was open as a gratuity to the public on Mondays and Saturdays, and permission to visit could be obtained from John Clarke, at the keeper's lodge. The menagerie, in addition to a giraffe, also had a collection of antelope, kangaroos, ostrich, gnus, emus and zebras. John Claudius Loudon, the Scottish botanist and garden designer, did not visit Sandpit Gate on one of the giraffe's 'state days'. Undeterred, he sneaked glimpses of the giraffe through the gaps in the wooden weatherboarding of the barn in which the giraffe was lodged. Loudon wrote, 'he appears thin, sickly, and very inferior to the giraffe of the Jardin des Plantes.' In truth, King George's little giraffe was something of a short straw. The Viceroy had intended to gift a giraffe each to both the French and the English, and had two giraffes bought to Cairo for that purpose. One of the giraffes was a young healthy female calf, and the other the sickly male. Such was the difference in the stature and state of the giraffes that the French and English consuls drew lots. The French giraffe lived for almost two decades after arriving in Paris in July 1827, but George's barely made two years. Whilst being bound and strapped onto a camel, George's giraffe had developed inflamed joints and become deformed with an odd gait. Although not a physically impressive creature, the giraffe was noted for his docile and gentle disposition and was said to more willingly lick the hands of the ladies than those of gentlemen. He was fed on a diet of milk, oats, hay, barley, split-beans and acacia shoots and was given regular walks in a paddock until he became confined though illness. The serious joint and neck problems necessitated the creation of a pulley and sling, as well as a frame-like device to keep the poor giraffe upright. In the same way that his mother had been strongly associated with her zebra, George IV featured in many satirical caricatures alongside his giraffe; one such caricature featured the giraffe being hoisted up with the pulley by a portly King George and his mistress, Lady Conyngham.

On 11 October 1829, the *Literary Gazette* reported, 'our interesting acquaintance the giraffe died.' George IV prepared to have his giraffe preserved for posterity in the galleries at Windsor Castle. The giraffe's bones were polished and articulated; a cast was taken of his body in order to create a wooden base over which the prepared giraffe skin could be stretched. King George IV died the following year, and his successor William IV did not care particularly for animals either at Sandpit Gate or at the Tower of London. Moreover, George IV had liked to spend money a little too liberally and yet had done little to endear the institution of the monarchy to the public.

At the end of the long eighteenth century the place of exotic animals in London, and more widely Britain, was quite different to that at the beginning. The foundation of the Zoological Society in 1822 and the opening of the London Zoological Gardens in 1828 had profound implications for the menageries and animal merchants that had developed in the city. Menageries feature prominently in late Georgian print, including Edward Bennett's *The Tower Menagerie: Comprising the Natural History of those Animals Contained in that Establishment* (1829); James Rennie's *The Menageries: Quadrupeds Described and Drawn from Living Subjects* (1831); and another work of Edward Bennett's, *The Gardens and Menagerie of the Zoological Society Delineated: Being Descriptions and Figures in Illustration of the Natural History of the Living Animals in the Society's Collection* (1831).

Replete with allegorical tales, anecdotes, biographies and citations of exotic animals both living and dead in collections, these printed natural histories reflect many of the traditions of the eighteenth century. But this proliferation of menageries in print was published at the same time as the closures of the Tower Menagerie and the Exeter 'Change Menagerie; and the opening of the new London Zoological Gardens.

The London Zoological Society was founded in 1826 by Sir Stamford Raffles, with the menagerie and gardens opening in 1828 on the northern edge of Regent's Park. The society had been established to promote the serious study of birds and animals as part of the new discipline of zoology, and it was seen as a marked improvement on vulgar menageries like that at the Exeter 'Change. The zoological gardens were at first open to fellows of the Zoological Society and their guests, and became a fashionable place for a promenade. It was not until 1847 that the zoological gardens were open to the public. This meant that, after the closure of the Exeter 'Change menagerie, London's best animal collection was a members-only club for almost two decades.

In 1829, the Exeter 'Change Menagerie was closed and the animal occupants moved by Edward Cross to temporary lodgings on the site of the present-day National Gallery, before he established his new Surrey Zoological Gardens in 1831 – a commercial venture intended to rival the Zoological Society. The 'Change had been the site of wild animal exhibitions and trade since at least the 1770s. Cross was moved in part due to the planned widening of the Strand, and in part on account of the nuisance or liability his menagerie was seen to have become, especially in the wake of the Chunee incident. Later, in 1834–5, the last animals in the Tower Menagerie left the Tower of London on the orders of the Duke of Wellington and joined the other animals that had earlier trickled into the collection of the London Zoological Society in 1831. The dispersal of the royal collection at the Tower and the donation of the private menagerie of William IV at Sandpit Gate to the Zoological Society were partially financially motivated but also politically expedient.

In the late 1820s and early 1830s, the formation of the menagerie of the Zoological Society at Regent's Park and the relocation of royal animals into it also coincided with a period of significant political instability and reform in Britain. Rising food prices and economic instability further motivated protest and revolutionary agitation alongside much broader demands for electoral reform. The zoological gardens that emerged from the early 1830s onwards were an important part of civic provincial middle-class life in cities such as Dublin (1831), Liverpool (1832), Manchester (1836) and Edinburgh (1839). They were established by subscribers or donors as part of learned societies or civic institutions; the animals within were given to the collection by donors. These new institutions corresponded to the increased confidence and political assertions of the provincial middle classes and their appeasement in the property provisions required for limited enfranchisement in the Reform Act of 1832. In 1831, as a prominent Tory opposed to Whig reform, the Duke of Wellington had been the target of angry protestors who gathered outside his Apsley House residence in London to break the windows. In Derby protestors stormed the city gaol; in Nottingham the Duke of Nottingham's residence was set ablaze; protest simmered across vast tracts of the West Country and in Birmingham and Manchester. The political crisis was heightened further in early May 1832 when protestors attempted to collapse the Bank of England by withdrawing private funds in gold. Against the revolutionary agitation, William IV donated his private menagerie to the Zoological Society in 1831, and as warden of the Tower of London the Duke of Wellington began to remove the animals to the ownership of the Zoological Society, improving the fortification and military capacity of the Tower in the process. Royal captive animals had become something of a political and financial liability.

This was not as sharp a disjuncture as one might imagine; these new institutions often inherited both the animals and attitudes of earlier menagerie proprietors, patrons and spectators. But in substantial ways the early zoological gardens were quite different in the specific claims to utility they made. The giraffe that had made it to Paris had flourished at the illustrious Jardin des Plantes, the French national botanical gardens, museum and menagerie. The sickly English giraffe that had been landed by Edward Cross had languished in a wooden barn at Sandpit Gate. Worse still, from the point of view of English pride and that of the new professional zoologists, the best collection of wild animals on view in London was Edward Cross's overtly commercial establishment on the Strand; housed in inadequate apartments above a busy thoroughfare. The menagerie at the Tower of London was, though not unimpressive, unfortunately housed in a fortress. The lack of a suitable national collection of exotic birds and animals had become conspicuous and embarrassing.

Both Rennie's and Bennett's works reveal some of the perceptions held by late Georgian naturalists and zoologists about the menageries of the previous generations. Rennie's *The Menageries* (1831) made a polemical argument for the proper 'Uses of the Menagerie'. A new defined national collection was a bold and necessary departure from the inadequate antecedents of the zoological garden. The value of a menagerie depended on its prudent arrangement, construction and regulation; in these matters it was clear that the Exeter 'Change was not up to scratch. Animals ought to be exhibited as far as was possible in their natural state. It was not possible, for example, to acquire 'an adequate notion of the kangaroo in a cage, but in a paddock its remarkable bound fixes our attention and curiosity'. Likewise lions and tigers in cramped dens did not exhibit their natural characteristics either: 'to put such an animal in a den is torture to him, and give false notions of his habits.'

In his condemnation of menageries Rennie employed two animals that have appeared throughout this book: the kangaroo and the elephant. In the case of the elephant Rennie dismissed the educational efficacy of menageries, claiming that spectators had, in actuality, formed precious few adequate notions of elephants in captivity. He contrasted the different quality of knowledge about elephant habits formed by the exhibition of elephants in menageries, on the London theatre stage and in a zoological garden like the Jardin des Plantes:

> Whatever interest we may feel in the sagacity which is already displayed by the elephants of our common English menageries, the wretched state of confinement in which so large an animal is kept prevents us from forming any adequate notions of many of its peculiarities. For this reason the most recent exhibition of the elephant in the theatre has contributed very much to remove some of the popular prejudices concerning the quadruped, and to induce correct ideas of its peculiar movements. We cannot, indeed, upon a stage, see the animal bound about as in a state of nature – roll with delight in the mud to produce a crust upon the body which should be impervious to its tormentors the flies – collect water in its trunk, to sprit over its parched skin – and browse upon the tall branches of trees which it reaches with its proboscis. We shall not see these peculiarities of its native condition, until we have a proper receptacle for the elephant in our national menagerie, the Zoological Gardens. Without imputing blame to those who exhibit the elephant in this country, there is great cruelty in shutting up in a miserable cage a creature who has such delight in liberty, and who is so obedient without being restrained.

Rennie, perhaps through graciousness rather than a lapse of memory, chose not to mention the shooting of Chunee. That incident was something of an elephant in the room, having occurred only a few years earlier. Instead he noted that an elephant in Atkin's Menagerie suffered in captivity. Rennie's elephant observations were salient indeed; as he wrote, a new paddock was in construction at the London Zoological Gardens to house the elephant from the Tower Menagerie, and another acquired by the Zoological Society. This paddock opened in August 1832, contained a water pool and trees and was many times larger than any earlier menagerie den.

In his introduction to the first delineation of the London Zoological Society's collection, Edward Bennett, Secretary of the

Zoological Society, proudly noted that both the gardens and the museum of the society had acquainted 'our countrymen' with animal creation, 'whether previously attached to zoology or indifferent to its allurements'. Such were the delights of a fashionable parade in the gardens that, in addition to 'more legitimate stimulants', it was not a matter of surprise that 'there should have arisen in the public mind a taste for zoological pursuits'.

In the early years, living animals from aristocratic menageries and the royal collection poured into the society's gardens and museum as donors sought recognition as patrons; George IV's stuffed giraffe was donated to the museum in 1831 instead of being installed at Windsor Castle; King William clearly did not want it around, gathering dust. The gardens also acquired animals from the sale of previous collections: the sale of the menagerie of the anatomist Joshua Brookes, for example, gave the society the opportunity to buy a vulture. Edward Cross too sold some of his animals to the Zoological Society. Of course, this cultural taste for exotic animals had deep roots, extending back into the eighteenth century; indeed, the new fashion for zoological pursuits was in many ways a recasting of an older Georgian interest in natural history and menageries. Moreover, Bennett chose to ignore the menagerie behaviours of visitors to the zoological gardens; they were not so much better behaved nor, in reality, did they ponder zoological matters more seriously than those who had crammed into the menageries and animal merchants on the Strand.

Yet, although a significant public existed for exotic animals in eighteenth-century London, the Zoological Garden in the early nineteenth century did much to promote the professionalization of zoology and wrest animal spectacles out of the grasp of showmen. As a national collection, albeit one initially restricted to a limited public of subscribed members and their guests, the London Zoological Gardens was a showpiece for natural history and zoology in Britain. Early nineteenth-century zoologists wanted to secure greater control over of the sources of their knowledge for the new discipline of zoology by establishing an integrated menagerie, museum and garden. The most prominent and vocal advocates of the new zoological gardens were, fittingly, two individuals with much invested in professional zoology: James Rennie, the first professor of zoology at King's College London, and Edward Bennett, the secretary of the Zoological Society between 1831 and 1836. Joshua Brookes, selling birds in Holborn in the 1760s, had called himself a 'zoologist'; this would have been anathema to men like Rennie and Bennett.

The world of the Georgian London animal merchants and menagerists, and that of the new London Zoological Society, however, overlapped in some unexpected ways. The life and expertise of Abraham Dee Bartlett, superintendent of the London Zoological Gardens, was a barometer of continuity and change. Edward Cross had been a friend of Bartlett's father, so as a young boy Bartlett was allowed to 'crawl about' the menagerie without paying for entry. He recalled playing with the young lions and 'other animals that were not likely to harm me': 'I have not the remotest recollection of seeing for the first time lions, tigers, elephant, or any other wild beasts, simply because I was almost from my birth among them.' He claimed to 'remember well' the shooting of Chunee in 1826, because he had been there as a 14-year-old boy. Cross would give Bartlett the dead bodies of those birds he had particularly loved feeding to take home, where he learnt to preserve them. It was these skills as a taxidermist that would bring him to the attention of the London Zoological Society. Bartlett himself considered those years crawling around the menagerie a formative time: 'the familiarity with wild beasts in my infancy has been of invaluable service to me.' In addition to serving as superintendent of the zoological gardens for almost 40 years, Bartlett also provided Queen Victoria with any number of birds that she required, taking care of them when she was absent. In his old age there could not have been many Londoners who remembered London's Georgian menageries, fewer still who had seen Chunee in person, and probably no other who had been there at his death. The passing of the Exeter 'Change back in 1829 had seemed, even to contemporaries, to mark the end of an era. The roars of the lions that had rumbled down the Strand and over the surrounding rooftops were to be heard no more:

> Famed Exeter 'Change is, alas! no more;
> Good bye to the Apes and all their Monkey tricks,
> With the Lion's and Tiger's melodious roar.'[1]

Note on Archives

Unless otherwise stated the archival material in this book can be found in the Lysons Collection, the John Johnson Collection of Printed Ephemera and the British Museum.

The Lysons Collection at the British Library and the John Johnson Collection of Printed Ephemera at the Bodleian Library, University of Oxford, are significant collections of eighteenth-century newspaper clippings, broadsides and handbills. The original Lysons folios are not available for public consultation but can be seen on microfilm (MC 20452, 26).

The British Museum's Department of Prints and Drawings has a substantial collection of eighteenth-century trade cards and handbills, including those of several bird and animal merchants.

The policy registers of the Sun Insurance Company must be consulted at the London Metropolitan Archives or Guildhall Library since they do not exist in digital format. These have been partially indexed by occupation, address or name but are best searched by policy number.

Notes

Prologue

1 The journals, diaries, notebooks, and collected ephemera of Richard Hall (1729–1801) are in the family collection of Mike Rendell, a descendant of Hall. Mike Rendell, *The Journal of a Georgian Gentleman: The Life and Times of Richard Hall, 1729–1801* (ebook, 2011).

2 The long eighteenth century is a more distended historical period than the chronological eighteenth century; this long century is usually bookended by the 'Glorious Revolution' of 1688 and the Reform Act of 1832. Other historians close the long eighteenth century with the end of the Napoleonic Wars in 1815, or on the death of King George IV in 1830. However defined, the long eighteenth century is an attempt to approach history in terms of the contours of cultural continuity and change beyond the chronological boundaries of 1701–1800. For the purposes of *The Georgian Menagerie* the long eighteenth century extends from the arrival of the 'strange and wonderful elephant' in London in 1675 to the closure of the Exeter 'Change Menagerie on the Strand in 1829.

Introduction

1 The United Provinces or Dutch Republic, now the Netherlands, was a confederation comprising seven provinces.

Part I: Trade

1 Advertisement for David Randal, bird merchant, *Flying Post*, 7–9 March 1704.

1 'Buy a Fine Singing Bird'

1 Trial of John Girle, July 1756 (t17560714-26), and Ordinary of Newgate's Account, September 1756 (OA17560920), Old Bailey Proceedings Online. Available at www.oldbaileyonline.org (accessed 20 February 2015). Hereafter OBP.

2 R. Campbell, *The London Tradesman: Being a Compendious View of All the Trades* (London: Gardner, 1747).

3 *Postboy*, London, 14 January 1701 (Burney Collection, British Library); 32 *Postman and the Historical Account*, London, 16 March 1706; *Tatler*, London, 26 December 1710 (Burney Collection, British Library).

4 Thomas Ward, *The Bird Fancier's Recreation, Including Choice Instructions for the Taking, Feeding, Breeding and Teaching of Them* (London: printed for T. Ward, 1735); *Daily Courant*, London, 17 January 1717 (Burney Collection, British Library).

5 Thomas Ward, *A Complete and Humorous Account of All the Remarkable Clubs and Societies in London and Westminster* (London: Wren, 1756).

6 Trial of John Marshall, December 1744 (t17441205-43), OBP.

7 William Ellis, *The Country Housewife's Family Companion* (London: James Hodges, 1750).

8 *Gazetter and New Daily Advertiser*, London, 15 May 1766; *Public Advertiser*, London, 24 December 1766; *Public Advertiser*, London, 28 September 1765 (Burney Collection, British Library).

9 Eighteenth-century spelling and punctuation has been rendered into modern British English. To avoid confusion, different contemporary spellings of the same name have been standardized: thus 'Brookes' and not the alternative spelling 'Brook'.

10 Joshua Brookes, zoologist, handbill (British Library, L.23.3 (48)). Lucy Agar Marshall, 'Samuel Marsden Brookes', *Wisconsin Magazine of History* lii (1968), pp. 51–9.

2 'To be Seen or Sold'

1   Anon., *Letters from an Irish Student in England to his Father in Ireland*, Vol. 1 (London: Lewis, 1809), pp. 149–50.

2   *Gazetter and New Daily Advertiser*, London, 21 February 1772 (Burney Collection, British Library).

3   John Cross Prob. 1776 and 1777, 11/1027 (Public Record Office, The National Archives).

4   Joshua Brookes, 20 February 1777, 380450, Sun Insurance Policy Register MS11936 (London Metropolitan Archives/Guildhall Library).

5   Joshua Brookes Prob. 11/1385 (Public Record Office, The National Archives).

6   From the collection of Jean Smiter, née Brookes: letter from R.H. Smith at Phillip Castang Ltd to Mr Brooks, 19 August 1955; letter from Phillip Castang to Mr Brookes, 1955; letter from R.H. Smith to *Country Life*, 7 February 1955.

7   Paul Brookes, Original Menagerie New Road, trade card, *c.*1810 (British Museum, Banks.14.3).

8   *The Annual Register 1816* (London: Baldrick and Cradock, 1817).

9   Trial of Mazarine Bell, 14 January 1807 (t18070114-100), OBP.

10 *Sporting Magazine*, London, xvi (1800), p. 128.

11 Gilbert Pidcock, 9 March 1803, 730081, Sun Insurance Policy Register MS11936 (London Metropolitan Archives/Guildhall Library).

12 Thomas Smith, *The Naturalist's Cabinet: Containing Interesting Sketches of Natural History* (London: Cundee, 1806).

13 Gilbert Pidcock, 27 July 1809, 832885, Sun Insurance Policy Register MS11936 (London Metropolitan Archives/Guildhall Library).

14 George Kendrick at 42 Piccadilly, 816235, Sun Insurance Policy Register MS11936 (London Metropolitan Archives/Guildhall Library).

15 William Granger, *The New Wonderful Museum*, Vol. II (London: Allen, 1804), p. 711; George Wilson, *The Eccentric Mirror: Reflecting a Faithful and Interesting Delineation of Male and Female Characters Ancient and Modern*, Vol. II (London: Cundee, 1807), p. 32. A woodcut of John Bobey with a short description appears in N. Burt,

Delineation of Curious Foreign Birds and Beasts, in their Natural Colours, which Are to be Seen Alive at the Great Room over Exeter Change (London: Jordan, 1791).

16 Stephen Polito Prob. 11/1556 (Public Record Office/The National Archives).

17 Edward Cross, 15 July 1818, 942859, and 30 August 1816, 921657 Sun Insurance Policy Register MS11936 (London Metropolitan Archives/Guildhall Library).

3 The Property of …

1 *Monthly Magazine* xxii/2 (London: Phillips, 1806), p. 130.

2 Eleazar Albin, *A Natural History of Birds, Vol. II* (London: Royal College of Physicians, 1731–8); George Edwards, *Gleanings of Natural History* (London: Royal College of Physicians, 1758).

3 Robert Townson, *Tracts and Observations in Natural History and Physiology* (London: White, 1799), p. 113.

4 Francis Griffin Stokes (ed.), *The Blecheley Diary of the Rev. William Cole 1765–67* (London: Constable, 1931); Francis Griffin Stokes (ed.), *A Journal of my journey to Paris in the year 1765, by the Rev. William Cole* (London: Constable, 1931); William Mortlock Palmer, *William Cole of Milton* (Cambridge: Galloway, 1935).

5 Thomas Hall, taxidermist, trade card (British Museum, D2.371); J. Thompson, taxidermist, trade card (British Museum, D2.359); John Chubb, taxidermist (British Museum, D2.395).

6 *Daily Courant*, London, 4 May 1706, 10 June 1707, 3 August 1716.

7 Trial of William Enoch, 13 April 1768 (t17680413-54), OBP; trial of Thomas Andrews, 22 April 1789 (t17890422-41), OBP.

8 *Sporting Magazine*, London, xxi (1803), p. 44.

9 *The Annual Register 1802* (London: Wilks, 1803), p. 455; *Spirit of the English Magazine* vi (Boston: Cotton, 1825), p. 390; Theodore Cook, *Eclipse and O'Kelly* (New York: Dutton, 1907).

10 *London Chronicle or Universal Evening Post*, 4–6 March 1762; *Kirby's Wonderful and Eccentric Museum, Vol. V* (London: Kirby, 1820), p. 28.

Part II: Ingredients

4 'Turtle Travels Far'

1 John Hawkesworth, *An Account of the Voyages Undertaken by the Order of his Present Majesty for Making Discoveries in the Southern Hemisphere ... Drawn up from the Journals which Were Kept by the Several Commanders and from the Papers of Sir Joseph Banks, Esq* (London: Cadell, 1773), p. 174.

2 Samuel Ward, *A Modern System of Natural History*, Vol. IX (London: Newbery, 1776).

3 *Morning Herald and Daily Advertiser*, 15 September 1783.

4 Charlotte Mason, *The Ladies' Assistant for Regulating and Supplying the Table* (London: Walter, 1787), p. 211. See also Hannah Glasse, *The Art of Cookery Made Plain and Easy* (London: Strahan, 1747).

5 John Timbs, *Anecdotes of the Clubs, Coffee Houses, and Taverns of the Metropolis in the 17th, 18th, and 19th Centuries*, Vol. I (London: Bentley, 1866).

6 'A Turtle Feast Described', *London Magazine or Gentleman's Intelligencer*, May 1755.

7 *Monthly Review or Literary Journal*, London, xxx (1764), p. 239. See also Sylvanus Urban, *The Gentleman's Magazine and Historical Chronicle for the Year 1773* (London: Newbery, 1773), p. 518.

8 William Kenrick, *The Widow'd Wife: A Comedy* (London: Davies, 1767), p. 36. See also *Supplement to Bell's British Theatre, Consisting of the Most Esteemed Farces and Entertainments now Performing on the British Stage*, Vol. IV (London: Bell, 1774), p. 361.

9 John Latimer, *Annals of Bristol in the Eighteenth Century* (London: Frome, 1893), p. 405.

10 Latimer, *Annals of Bristol in the Eighteenth Century*, p. 323.

11 The dining register of the club is annotated and cited in Archibald Geikie, *Annals of the Royal Society Dining Club: The Record of a London Dining Club* (London: Macmillan, 1917).

12 George Selwyn's turtle letters can be found in: John Henage, *George Selwyn and his Contemporaries: With Memoirs and Notes* (London: Bentley, 1843).

13 Horace Walpole, *Private Correspondence of Horace Walpole, Earl of Orford, Vol. I, 1735–1756* (London: Rodwell, 1820).

14 Ben Rogers, *Beef and Liberty: Roast Beef, John Bull, and the English Patriots* (London: Chatto and Windus, 2003).

15 Margaret Lenta, *The Cape Diaries of Lady Anne Barnard, Vol. I, 1799–1800* (Cape Town: Van Riebeeck Society, 1999).

5 Bear Grease for your Powdered Wig

1 Craig Horner (ed.), *The Diary of Edmund Harrold, Wigmaker of Manchester, 1712–15* (Aldershot: Ashgate, 2008).

2 *Bell's Monthly Compendium of Advertisements*, October 1807.

3 John George Wood, *Sketches of Animal Life* (London: Routledge, 1855).

4 *Sunday Reformer and Universal Register*, 30 June 1793; *Parker's General Advertiser*, 26 October 1782; *Dublin Mercury*, 14 October 1769; *An Ode to Mr. Lewis Hendrie, Principal Bear-Killer in the Metropolis of England and Comb-Maker in Ordinary to His Majesty* (London: Bladon, 1783).

5 Anon., *The New London Toilet* (London: Richardson, 1778).

6 James Rennie, *The Art of Preserving the Hair* (London: Prowett, 1826).

7 Clark Erskine, *Chatterbox* (Boston: Erste and Co., 1906); *Annual Register 1806* (London: Longman, 1808), p. 456; *Sporting Magazine*, London, xxix (1807).

8 'The Posthumous Letters of Charles Edwards Esq', *Blackwood's Edinburgh Magazine*, i New Series (1826), p. 22.

9 Cyrus Jay, *The Law: What I Have Seen, What I Have Heard, What I Have Known* (London: Tinsley, 1868).

10 'The Rival Bears', *The Spirit of the Public Journals* (London: Sherwood, 1825), p. 438.

11 'Perfumes and Razors', *The Spirit of the Public Journals, Vol. X, 1806* (London: Ridgway, 1807), pp. 316–17.

12 *London Monthly Magazine or the Gentleman's Monthly Intelligencer for the Year 1763* (London: Baldwin, 1763), p. 32.

6 'The Product of the Civet's Posteriors'

1 William Fordyce Mavor, *Natural History for the Use of Schools* (London: Phillips, 1800).

2 Pierre Pomet, *A Complete History of Drugs* (London: Bonwicke, 1684); John Hill, *A History of Materia Medica* (London: Longman, 1751).

3 William Salmon, *Pharmacopœia Londinensis: Or the New London Dispensatory* (London: Nicholson, 1702); John Martin, *A Treatment of the Venereal Disease* (London: Marten, 1711); Leonard Sowerby, *The Ladies Dispensary: Or Every Woman her own Physician* (London: Hodges, 1740).

4 Charles Lillie, advertisement, *The Spectator* ii (1747), p. 331; Charles Lillie, trade card, 1736 (British Museum 1910.1208.14).

5 Charles Lillie, *The British Perfumer*, ed. Colin Mackenzie (London: Souter, 1822).

6 Thomas Phillips, *A Journal of a Voyage Made in the* Hannibal *of London, Ann. 1693, 1694* (London: Phillips, 1732); Paula Backscheider, *Daniel Defoe: His Life* (Baltimore: Johns Hopkins University Press, 1989), pp. 55–8.

7 Robert Boyle, *The Philosophical Works of Robert Boyle*, ed. Peter Shaw (London: Longman, 1738).

8 Alexander Pope, *The Works of Alexander Pope Esq.* (London: Dodsley, 1738); Anon., *The Lover's Pacquet: Or the Marriage Miscellany* (London: Reynolds, 1733); William Cowper, *Poems by William Cowper Esq.* (London: Johnson, 1787); 'Luxury Pernicious to Persons and States', *London Magazine or Gentleman's Monthly Intelligencer for 1749, Vol. XVII* (London: Baldwin, 1749), p. 537.

9 Oliver Goldsmith, *A History of Earth and Animated Nature* (London: Nourse, 1774); Richard Warren, handbill, 1777, available at Eighteenth Century Collections Online, http://quod.lib.umich.edu/e/ecco (accessed 20 February 2015).

Part III: Crowds

1 George Arnold, *Robert Pocock: The Gravesend Historian, Naturalist, Antiquarian, Botanist, and Printer* (London: Sampson, 1883).

7 Ladies and Gentlemen

1 *Jackson's Oxford Journal*, 23 January 1779 (Bodleian Library N.G.A. Oxon a.4); Gilbert Pidcock, *The History, and Anatomical Description of a Cassowar, from the Isle of Java, in the East-Indies, the Greatest Rarity Now in Europe* (Bury: Green, 1778); N. Burt, *Delineation of Curious Foreign Birds and Beasts, in their Natural Colours, which Are to be Seen Alive at the Great Room over Exeter Change* (London: Jordan, 1791); T. Garner, *A Brief Description of the Principal Foreign Animals and Birds Now Exhibiting over Exeter Exchange* (London: T. Burton, printed for Gilbert Pidcock, 1800).

2 Anon., *A True and Perfect Description of the Strange and Wonderful Elephant Sent from the East Indies* (London: Sumpter, 1675); Allen Moulin, *An Anatomical Account of the Elephant Accidentally Burnt in Dublin* (London: Smith, 1682); Patrick Blair, *Osteographica Elephantina: Or a Full and Exact Account of the Bones of an Elephant which Died near Dundee* (London, 1713); William Stukeley, *An Essay towards the Anatomy of the Elephant* (London: printed for the author, 1723).

3 Michiel Roscam Abbing, 'So Een Wunder heeft men hier nooijt gesien' de indische vrouwtjesolifant (1678/80–1706) van bartel verhagen', *Jaabock 106, Amstelodamum* (2014), pp. 12–39.

4 Anon., *A Letter from the Elephant to the People of England* (London: Sumpter, 1764); Anon., *The Elephant's Speech to the Citizens and Countrymen of England* (London, 1675); 'The Great White Elephant, Alive, is to be Seen in this Town', single-sheet folio (John Johnson Collection, Bodleian Library). John Booth, *A Century of Theatrical History, 1816–1916* (London: Stead, 1917); James Rennie, *The Menageries: Quadrupeds Described and Drawn from Living Subjects* (London: Wells, 1831).

5 Arthur Phillip, *The Voyage of Governor Phillip to Botany Bay* (London: Stockdale, 1789); John White, *Journal of a Voyage to New South Wales* (London: Debretts, 1790).

6 'Kangaroo at Pidcock's Menagerie', *Morning Chronicle*, 16 December 1803; 'Aristocracy at Pidcock's Menagerie', *Morning Chronicle*, 18 April 1801. *The Asiatic Journal and Monthly Register for British India and its Dependencies* ix (January–June 1820), p. 40. James Hardy Vaux and Barron Field, *Memoirs of James Hardy Vaux, Written by Himself in Two Volumes, Vol. I* (London: Clowdon, 1819), p. 20. William Fordyce Mavor, *Natural History for the Use of Schools* (London: Phillips, 1800), pp. 101–3; Marc Auguste Pictet, *Voyage de trois mois en Angleterre, en Ecosse, et en Irlande pendant l'été de l'an IX* (Paris: Manget, 1802); Auguste von Kotzebue, *Voyages through Italy in the Years 1804 and 1805, Vol. IV* (London: Phillips, 1806), p. 305. *The Monthly Review; or Literary Journal, Enlarged: From January to April inclusive, Vol. XLIII* (London: Straban, 1804), p. 498.

8 Bitten, Crushed and Maimed

1 John Moffat, *The History of the Town of Malmesbury* (Tetbury: Goodwyn, 1805).

2 John Evelyn, *The Diary of John Evelyn*, ed. Guy de la Bédoyère (Woodbridge: Boydell, 2004), pp. 265 and 86.

3 Edward Ward, *The London Spy Compleat in Eighteen Parts. Being the First Volume of the Writings of Mr. Edward Ward* (London: Bettesworth, 1718), pp. 303–7. David Henry, *An Historical Account of the Curiosities of London and Westminster* (London: Newbery, 1769), pp. 16–19; William Bingley, *Animal Biography, Vols I and II* (London: Phillips, 1803).

4 Thomas Wright, *The Life of William Cowper* (London: Haskell House, 1892).

5 Nathaniel Wheaton, *A Journal of Residence during Several Months in London* (Hartford: Huntington, 1830).

6 *The Literary Panorama, Being a Compendium of National Papers and Parliamentary Reports, Vol. XI* (London: Baylis, 1812), p. 167; *The*

Annual Register, or a View of the History, Politics, and Literature for the Year 1811 (London: Longman, 1812), p. 99.

7   Charles Orpen, *Anecdotes of the Deaf and Dumb* (London: Timms, 1836), pp. 413–15, 432–4.

8   Cornelius Nicholson, *The Annals of Kendal: Being an Historical and Descriptive Account of Kendal and the Neighbourhood with Biographical Sketches of Many Eminent Personages Connected with the Town* (London: Whitaker, 1861).

9   'To be Seen a Young Crocodile', July 1789, advertisement (Lysons Collection, British Library RBMRR: 1881.b.6; 1889.b.5; 1889.e.6); 'The Boa Constrictors, or Great Serpents Alive', advertisement, *Caledonia Mercury*, 17 August 1816 (Burney Collection, British Library); Edward Bennett, *The Tower Menagerie: Comprising the Natural History of the Animals Contained in that Establishment with the Anecdotes of their Characters and History* (Chiswick: Whittingham, 1829).

10   'The Case of a Man who Died in Consequence of the Bite of a Rattlesnake: With an Account of the Effects Produced by the Poison', *Philosophical Transactions of the Royal Society of London* c (1810), pp. 75–88; *European Magazine and London Review* lvi (1809), p. 430.

11   *Lancaster Gazette*, 28 July 1827; 'Fearful Accident: Four Lives Lost!', broadside, *c*.1834 (National Library of Scotland F.3.a.13(115)).

12   James Parsons, 'A Letter from Dr. Parsons to Martin Folkes, Esq. President of the Royal Society, Containing the Natural History of the Rhinoceros', *Philosophical Transactions of the Royal Society of London* xlii (1742), p. 532–41; Everard Home, 'On a New Species of Rhinoceros Found in the Interior of Africa', *Philosophical Transactions of the Royal Society for the Year 1821* (London, Nicol, 1822), p. 44.

13   'Death Caused by Elephant', *Examiner*, 7 November 1825; 'The Obituary of John Tietjen', *Gentleman's Magazine* xcv (1825), p. 475; John Sykes, *Local Records: Or, Historical Register of Remarkable Events which Have Occurred in Northumberland and Durham, Newcastle Upon Tyne, and Berwick Upon Tweed* (Newcastle: Sykes, 1833).

9 Sweet Camel's Breath

1 'Dromedary and Camel', advertisement, 26 January 1758; *Mist's Journal*, cutting, 5 April 1758 (Lysons Collection, British Library, microfilm MC20452, frames 9 and 8e); *London Magazine or Gentleman's Intelligencer for 1758, Vol. XXVII* (London: Baldwin, 1758), pp. 247–9.

2 'A Live Boos Potamus, or the River Cow of Egypt, from the Banks of the Nile', advertisement, 1799 (Lysons Collection, British Library, microfilm MC20452, frame 26).

3 *An Excellent New Ballad on the South-Sea Dog-Fish that was Shewn on the River Thames in July 1725* (London: Lewis, 1726).

4 Edward Ward, *The London Spy Compleat in Eighteen Parts: Being the First Volume of the Writings of Mr. Edward Ward* (London: Bettesworth, 1718).

5 'The Piked Whale', advertisement, 29 January 1787 (Lysons Collection, British Library, microfilm MC20452, frame 53).

6 Thomas Smith, *The Naturalist's Cabinet: Containing Interesting Sketches of Animal History* (London: Cundee, 1806); William Bingley, *Animal Biography, Vol. I* (London: Phillips, 1803); T. Garner, *A Brief Description of the Principal Foreign Animals and Birds Now Exhibiting over Exeter Exchange* (London: T. Burton, printed for Gilbert Pidcock, 1800).

7 Robert Jameson, 'Journal of a Voyage from Leith to London, 1793' (University of Edinburgh Special Collections, GB 0237 Robert Jameson Dc.5.34).

8 'Mr Patterson's Ostrich, 37 Strand', handbill, *c*.1795 (Bodleian Library ESTC, T187106).

9 'Monthly Memoranda of Natural History', *Scots Magazine for 1812, Vol. LXXIV* (Edinburgh: Constable, 1812), pp. 652 and 812; 'Letter from Jameson to Scoresby', in Tom and Cordelia Stamp, *William Scoresby, Arctic Scientist* (Whitby: Caedmon, 1976), p. 47.

10 Exotic Estates

1   Edmond Fitzmaurice, *Life of William, Earl of Shelburne, Afterwards First Marquess of Lansdowne, with Extracts from his Papers and Correspondence*, 2 vols (London: Macmillan, 1912); John Bowring, *The Works of Jeremy Bentham, Part XIX, Memoirs of Bentham* (Edinburgh: Tait, 1842).

2   Montagu Pennington, *A Series of Letters between Mrs. Elizabeth Carter and Miss Catherine Talbot from the Year 1744 to 1770* (London: Rivington, 1809); William Hayley, *The Life, and Posthumous Writings of William Cowper Esq.* (Chichester: Johnson, 1806).

3   Matthew Montagu, *The Letters of Mrs E. Montagu with Some of the Letters of her Correspondents, Vol. II* (Boston: Wells & Lilly, 1825).

4   Henry Foster to the Duke of Richmond in 1730, cited in Rosemary Baird, *Goodwood: Art and Architecture, Sport and Family* (London: Francis Lincoln, 2007), p. 63.

5   John Claudius Loudon, *The Landscape Gardening and Landscape Architecture of the Late Humphrey Repton Esq.* (London: Longman, 1840).

11 John Bull and 'Happy Britain'

1   *Sporting Magazine*, London, xxx (1808), p. 136; Thomas Smith, *The Naturalist's Cabinet: Containing Interesting Sketches of Animal History* (London: Cundee, 1806).

2   Samuel Ward, *A Modern System of Natural History, Vol. V* (London: Newbery, 1775), p. 91.

3   Pidcock advertisements in *London Packet*, 24 January 1800, and *Observer*, London, 2 March 1800; William Bullock, *A Companion to the London Museum* (London: Rowland, 1813).

4   Gilbert White, *The Natural History and Antiquities of Selborne* (London: White and Son, 1789), p. 80.

5   William Stukeley, *Of the Spleen, its Description and History … to which is Added an Essay towards the Anatomy of the Elephant* (London: printed for the author, 1723), p. 91; *Monthly Review or*

Literary Journal, May–August 1819, p. 61; letter from Mary Lamb to Elizabeth Betham, 2 November 1814, in Anne Gilchrist, *Mary Lamb* (Boston: Roberts, 1883), p. 264.

12 Llama's Spit, a Pot of Barclay's Entire and Elephant Chops

1 John Thomas Smith, *A Book for a Rainy Day: Or, Recollections of the Last Sixty-Six Years* (London: Bentley, 1845); John Thomas Smith, *A Book for a Rainy Day: Or, Recollections of the Years 1766–1833* (London: Bentley, 1861).

2 James Hall, *Travels in Scotland, Vol. II* (London: Johnson, 1807).

3 Benjamin Silliman, *A Journal of Travels in England, Holland, and Scotland and of Two Passages across the Atlantic in 1805 and 1806, Vol. I* (New York: Bruce, 1810).

4 'The Elephant as he Lay Dead at the Exeter Change', in William Hone, *The Everyday Book: Or Everlasting Calendar, Vol. II, Part II* (London: Hunt and Clark, 1827), pp. 321–36.

13 Under the Knife

1 Anon., *A True and Perfect Description of the Strange and Wonderful Elephant Sent from the East Indies* (London: Sumpter, 1675).

2 Allen Moulin, *An Anatomical Account of the Elephant Accidentally Burnt in Dublin* (London: Smith, 1682).

3 Patrick Blair, *Osteographica Elephantina: Or a Full and Exact Account of the Bones of an Elephant which Died near Dundee* (London, 1713); William Stukeley, *Of the Spleen, its Description and History … to which is Added an Essay towards the Anatomy of the Elephant* (London: printed for the author, 1723).

4 Joseph Adams, *Memoirs of the Life and Doctrines of the Late John Hunter* (London: Thorne, 1817); James Palmer, *The Works of John Hunter, Vol. IV* (London: Longman, 1837).

5 Henry Angelo, *Reminiscences of Henry Angelo, with Memoirs of his Late Father and Friends* (London: Colburn, 1828).

Part IV: Humour

14 Electric Desire

1   Lucretia Lovejoy, *An Elegy on the Lamented Death of the Electrical Eel, or Gymnotus Electricus* (London: Fielding and Walker, 1777–9).

2   Edward Bancroft, *An Essay on the Natural History of Guiana in South America* (London: Becket, 1769); 'An Account of the Gymnotus Electricus or Electrical Eel in a Letter from Alexander Garden, M.D.F.R.S. to John Ellis, Esq; F.R.S.', in *The Annual Register 1775*, 4th edn (London: n.p., 1783), pp. 87–92; Hugh Williamson, 'Experiments and Observations on the Gymnotus Electricus, or Electrical Eel by Hugh Williamson, M.D. Communicated by John Walsh, Esq, F.R.S.', *Philosophical Transactions: Proceedings of the Royal Society of London* lxv (1775), pp. 94–101.

3   Tiberius Cavallo, *A Complete Treatise on Electricity, in Theory and Practice; with Original Experiments, Vol. II* (London: Dilly, 1786). For some of Baker's electric eel advertisements see *Gazetteer and New Daily Advertiser*, London, 23 November 1776, 27 November 1776, 14 December 1776 (Burney Collection, British Library).

4   *The Torpedo, a Poem to the Electrical Eel. Addressed to Mr. Hunter, Surgeon* (London: n.p., 1777); James Perry, *The Electrical Eel; Or Gymnotus Electricus. Inscribed to the Members of the Royal Society, by Adam Strong, Naturalist* (London: Bew, 1777); Anon., *The Inamorato: Addressed to the Author of The Electrical Eel by a Lady* (London: Bew, 1777); Anon., *The Old Serpents Reply to the Electrical Eel* (London: Smith, 1777); Nathan Bailey, *An Universal Etymological English Dictionary* (London: Bailey, 1775).

5   Alexander Pettit and Patrick Spedding (eds), *Eighteenth-Century British Erotica* (London: Pickering & Chatto, 2002); Lyn Hunt (ed.), *The Invention of Pornography, 1500–1800: Obscenity and the Origins of Modernity* (New York: Zone Books, 1993).

15 The Queen's Ass

1   François Marie Arouet de Voltaire, *A Letter from Mr. Voltaire to Mr. Jean Jacques Rousseau* (London: Payne, 1766), p. 35.

2   'Some Account of the Zebra, or Painted African Ass, Lately Brought over and Presented to the Queen', *London Magazine and Gentleman's Intelligencer*, 31 July 1762; 'Extracts from the Orders Given to the Third Regiment of Guards', *Lloyds Evening Post*, 13 April 1762; *Gazetteer and New Daily Advertiser*, 27 June 1766; *Middlesex Journal or Chronicle of Liberty*, 21 September 1769.

3   Henry Howard, 'The Queen's Ass: A New Humorous Allegorical Song', broadside, London, 1762 (British Museum 0808.4198). Other zebra broadsheet satires are 'With a Fool's Head at the Tail: The Other Side of the Zebray' (1762); 'The Real Ass' (1762); 'The King's Ass' (1762); 'Zebra Rescued, or a Bridle for the Ass' (1762).

4   *Jackson's Oxford Journal*, 11 April 1772.

5   William Mason to Horace Walpole, 2 June 1773, in Warren Hunting Smith and George Lam (eds), *Horace Walpole's Correspondence, 1756–1799* (New Haven: Yale University Press, 1955), pp. 90–1.

6   Anon., *The Alphabetical Drawing Book and Pictorial History of Quadrupeds* (New York: Wiley and Putnam, 1847); William Bingley, *Animal Biography or Popular Zoology* (London: Rivington, 1820); Oliver Goldsmith, *An Abridgement of Dr. Goldsmith's Natural History of Beasts and Birds* (London: Whittingham, 1807); William Nicholson, *The British Encyclopedia; or Dictionary of Arts and Sciences* (London: Longman, 1809).

7   John Watkins, *Memoirs of Her Most Excellent Majesty Sophia Charlotte, Queen of Britain* (London: Colburn, 1819).

16 The Love Birds

1   John Marchant, *Puerilia: Or, Amusements for the Young. Consisting of a Collection of Songs, Adapted to the Fancies & Capacities of those of Tender Years* (London: Stevens, 1751); *The Tatler; Or Lucubrations of Isaac Bickerstaff* (London: Bathurst et al., 1707).

2 Joseph Highmore, 'To Miss S— H—, Desiring her Parrot Might be the Subject of a Poem', *Gentleman's Magazine* xx (April 1750), pp. 180–1.

3 Edward Wicksteed, *The Young Gentleman and Lady Instructed in Such Principles of Politeness*, Vol. II (London: Wicksteed, 1747).

4 John Moxton, *The Most Agreeable Companion; Or, a Choice Collection of Detached and Most Approved Pieces in Prose and Verse* (Leeds: Wright, 1782).

5 'The Lover and his Parrot', *London Magazine*, December 1732.

6 Ann Sheldon, *Authentic and Interesting Memoirs of Miss Ann Sheldon* (London: printed for the authoress, 1787).

7 Augusta Llanover, *The Autobiography and Letters of Mary Granville, Mrs. Delaney*, Vol. I (London: Bentley, 1862).

8 Samuel Johnson, *The Rambler and the Idler (1751–1752)* (London: Jones and Co., 1826).

9 Letter from Samuel Johnson to Hester Piozzi, 19 June 1775, Hester Lynch Piozzi, *Letters to and From Samuel Johnson*, Vol. I (Dublin: Moncrieffe, 1788), p. 178.

10 William Kendrick, *The Whole Duty of Woman* (London: n.p., 1753); Anon., *The Polite Lady; or a Course of Female Education* (London: Newbery, 1769).

11 Letter from Elizabeth Montagu to Mrs Donellan, 3 March 1744, in Matthew Montagu, *The Letters of Mrs E. Montagu with Some of the Letters of her Correspondents*, Vol. II (Boston: Wells & Lilly, 1825).

12 Tobias Smollett, *Travels through France and Italy* (first published London, 1766; 1979 edition edited by Frank Felsenstein, Oxford: Oxford University Press).

13 George Edwards, *A Natural History of Uncommon Birds*, Vol. II (London: printed for the author, 1747).

14 Thomas Hayes, *Portraits of the Rare and Curious Birds with their Descriptions from the Menagery of Osterley Park in the County of Middlesex*, Vol. I (London: W. Bulmer & Co., 1794–9); Daniel Lysons, *The Environs of London: Being an Historical Account of the Towns, Villages, and Hamlets within Twelve Miles of that Capital* (London: Cadell, 1796).

Epilogue

1 Edward Turner Bennett, *The Tower Menagerie: Comprising the Natural History of the Animals Contained in that Establishment with Anecdotes of their Characters and History* (Chiswick: Whittingham, 1829); Edward Turner Bennett, *The Gardens and Menagerie of the Zoological Society Delineated* (Chiswick: Whittingham, 1831); James Rennie, *The Menageries: Quadrupeds Described and Drawn from Living Subjects* (London: Wells, 1831); Edward Bartlett, *Wild Animals in Captivity: Being an Account of the Habits, Food, Management, and Treatment of the Beasts and Birds at the Zoo with Reminiscences and Anecdotes by A.D. Bartlett* (London: Chapman, 1899); Thomas Hudson, *Comic Songs by Thomas Hudson: Collection the Thirteenth* (London: Hudson, 1832), p. 4.

Further Reading

Exotic Animals in Eighteenth-Century Europe

Ash, Mitchell (ed.), *Mensch, Tier und Zoo: Der Tiergarten Schönbrunn im internationalen Vergleich vom 18. Jahrhundert bis heute* (Vienna: Böhlau, 2008).

Burkhardt, Richard, 'The Leopard in the Garden: Life in Close Quarters at the Museum d'Histoire Naturelle', *Isis* xcviii/4 (2007), pp. 675–94.

Donald, Diana, *Picturing Animals in Britain 1750–1850* (New Haven: Yale University Press, 2007).

Festing, Sally, 'Menagerie and the Landscape Garden', *Journal of Garden History* viii (1988), pp. 104–17.

Meisen, Lydia, *Die Charakterisierung der Tiere in Buffon's Histoire naturelle* (Würzburg: Königshausen & Neumann, 2007).

Morton, Mary (ed.), *Oudry's Painted Menagerie: Portraits of Exotic Animals in Eighteenth-Century Europe* (Los Angeles: Getty, 2007).

Palmeri, Frank (ed.), *Humans and Other Animals in Eighteenth-Century British Culture: Representations, Hybridity, Ethics* (Aldershot: Ashgate, 2006).

Ridley, Glynis, *Clara's Grand Tour: Travels with a Rhinoceros in Eighteenth-Century Europe* (London: Atlantic Books, 2005).

Rieke-Müller, Annelore and Lothar Dittrich (eds), *Unterwegs mit wilden Tieren: Wandermenagerien zwischen Belehrung und Kommerz 1750–1850* (Marburg: Basilisken Press, 1999).

Robbins, Louise, *Elephant Slaves and Pampered Parrots: Exotic Animals in Eighteenth-Century Paris* (Baltimore: Johns Hopkins University Press, 2002).

Senior, Matthew (ed.), *A Cultural History of Animals in the Age of Enlightenment* (London: Berg, 2007).

Tague, Ingrid, *Animal Companions: Pets and Social Change in Eighteenth-Century Britain* (University Park: Penn State University Press, 2015).

Eighteenth-Century Exotic Animal Fiction

Klinkenborg, Veryln, *Timothy's Book: Notes of an English Country Tortoise* (London: Portobello Books, 2007).

Nicholson, Christopher, *The Elephant Keeper* (London: HarperCollins, 2010).

Other Exotic Animal Histories (Britain)

Cowie, Helen, *Exhibiting Animals in Nineteenth-Century Britain: Empathy, Education, and Entertainment* (London: Palgrave Macmillan, 2014).

Hahn, Daniel, *The Tower Menagerie* (London: Simon and Schuster, 2003).

Simons, John, *Rossetti's Wombat: Pre-Raphaelites and Australian Animals in Victorian London* (London: Middlesex University Press: 1994).

——— *The Tiger that Swallowed the Boy: Exotic Animals in Victorian England* (Farringdon: Libri, 2012).

Velten, Hannah, *Beastly London: A History of Animals in the City* (London: Reaktion, 2013).

Other Exotic Animal Histories (Europe and Elsewhere)

Alberti, Samuel (ed.), *The Afterlife of Animals: A Museum Menagerie* (Charlottesville: University of Virginia Press, 2011).

Allin, Mark, *Zarafa: The True Story of Giraffe's Journey from the Plains of Africa to the Heart of Post-Napoleonic France* (London: Headline, 1998).

Baratay, Eric and Elisabeth Hardouin-Fugier, *Zoo: A History of Zoological Gardens in the West* (London: Reaktion, 2004).

Deiss, William and Robert Hoage (eds), *New Worlds, New Animals: From Menagerie to Zoological Garden* (Baltimore: Johns Hopkins University Press, 1996).

Dittrich, Lothar and Annelore Rieke-Müller, *Die Kulturgeschichte des Zoos* (Berlin: Verlag für Wissenschaft und Bildung, 2001).

Dittrich, Lothar and Annelore Rieke-Müller, *Der Löwe brüllt nebenan: die Gründung Zoologischer Gärten im deutschsprachigen Raum 1833–1869* (Vienna: Böhlau, 1998).

Fudge, Erica, *Renaissance Beasts: Of Animals, Humans, and Other Wonderful Creatures* (Chicago: University of Illinois Press, 2002).

Kalof, Linda, *Looking at Animals in Human History* (London: Reaktion, 2007).

Kete, Kathleen, *The Beast in the Boudoir: Pet-Keeping in Nineteenth-Century Paris* (Berkeley: University of California Press, 1994).

——— (ed.), *A Cultural History of Animals in the Age of Empire* (London: Berg, 2007).

Nichols, Henry, *Lonesome George: The Life and Loves of a Conservation Icon* (London: Macmillan, 2006).

———*The Way of the Panda: The Curious History of China's Political Animal* (London: Profile Books, 2010).

Rothfels, Nigel (ed.), *Representing Animals* (Bloomington: University of Indiana Press, 2002).

——— *Savages and Beasts: The Birth of the Modern Zoo* (Baltimore: Johns Hopkins University Press, 2002).

Sleigh, Charlotte, *Six Legs Better: A Cultural History of Myrmecology* (Baltimore: Johns Hopkins University Press, 2007).

Smith, Pamela and Paula Findlen (eds), *Merchants and Marvels: Commerce, Science and Art in Early Modern Europe* (London: Routledge, 2002).

Walker-Meikle, Kathleen, *Medieval Pets* (Woodbridge: Boydell Press, 2012).

Eighteenth-Century Cultural and Social Histories

Alltick, Richard, *The Shows of London* (Cambridge: Belknap, 1978).

Berg, Maxine, *Luxury and Pleasure in Eighteenth-Century Britain* (Oxford: Oxford University Press, 2005).

Brewer, John, *Pleasures of the Imagination: English Culture in the Eighteenth Century* (Chicago: University of Chicago Press, 1997).

Crowley, John, *The Invention of Comfort: Sensibilities and Design in Early Modern Britain and Early America* (Baltimore: Johns Hopkins University Press, 2001).

Daston, Lorraine and Katharine Park, *Wonders and the Order of Nature, 1150–1750* (London: Zone, 1998).

Fergus, Jan, *Provincial Readers in Eighteenth-Century England* (Oxford: Oxford University Press, 2006).

Gascoigne, John, *Joseph Banks and the English Enlightenment: Useful Knowledge and Polite Culture* (Cambridge: Cambridge University Press, 2003).

Gatrell, Vic, *City of Laughter: Sex and Satire in Eighteenth-Century London* (London: Atlantic Books, 2006).

Golinski, Jan, *British Weather and the Climate of Enlightenment* (Chicago: Chicago University Press, 2007).

Greg, Hannah, *The Beau Monde: Fashionable Society in Georgian London* (Oxford: Oxford University Press, 2013).

Inglis, Lucy, *Georgian London: Into the Streets* (Penguin: London, 2014).

Langford, Paul, *A Polite and Commercial People: England, 1727–1783* (Oxford: Oxford University Press, 1992).

Moore, Wendy, *The Knife Man: The Extraordinary Life and Times of John Hunter* (London: Bantam Press, 2005).

Ogborn, Miles and Charles Withers (eds), *Georgian Geographies: Essays on Space, Place, and Landscape in the Eighteenth Century* (Manchester: Manchester University Press, 2004).

O'Gorman, Frank, *The Long Eighteenth Century, 1688–1832* (New York: Arnold, 1997).

Rogers, Ben, *Beef and Liberty: Roast Beef, John Bull, and the English Patriots* (London: Chatto and Windus, 2003).

Styles, John and Amanda Vickery (eds), *Gender, Taste, and Material Culture in Britain and North America 1700–1830* (New Haven: Yale University Press, 2007).

Vickery, Amanda, *Behind Closed Doors: At Home in Georgian England* (New Haven: Yale University Press, 2010).

——— *The Gentleman's Daughter: Women's Lives in Georgian England* (New Haven: Yale University Press, 2003).

Index